INTERNATIONAL SERIES OF MONOGRAPHS IN
NATURAL PHILOSOPHY

GENERAL EDITOR: D. TER HAAR

VOLUME 69

THE NEBULAR VARIABLES

THE NEBULAR VARIABLES

By

JOHN S. GLASBY, B.SC., F.R.A.S.

PERGAMON PRESS

OXFORD · NEW YORK · TORONTO · SYDNEY

Pergamon Press Ltd., Headington Hill Hall, Oxford OX3 0BW

Pergamon Press Inc., Maxwell House, Fairview Park, Elmsford, New York 10523

Pergamon of Canada Ltd., 207 Queen's Quay West, Toronto 1

Pergamon Press (Aust.) Pty. Ltd., 19a Boundary Street, Rushcutters Bay, N.S.W. 2011, Australia

First edition 1974

Library of Congress Cataloging in Publication Data

Glasby, John Stephen.

The nebular variables.

(International series of monographs in natural philosophy, v. 69)

Includes bibliographies.

1. Stars, Variable. 2. Nebulae. I. Title.

QB835.G528 1974   523.8′44   74–3354

ISBN 0–08–017949–5

Printed in Great Britain by A. Wheaton & Co., Exeter

Contents

List of Plates

Foreword

MANY variable stars are known which are classed as irregular because of the almost complete lack of periodicity in their light variations. Among these we find a large number that are associated with nebulosity. These are the nebular variables which, because of their physical connection with either bright or dark nebulae, are among the most interesting and important of all the variable stars.

The study of their complex light fluctuations and spectral characteristics, coupled in many cases with those of the surrounding nebulosity, has provided astronomers with a vast amount of data concerning the pre-main sequence evolution of such stars.

In the main, we find that the variations in their brightness are entirely unpredictable. Nevertheless, it is possible to subdivide the nebular variables into certain fairly well-defined classes. Originally, two such groups were recognized; those typified by RW Aurigae and those found in large numbers within the Orion complex and named after the prototype star, T Orionis. As more data accumulated, a third group was added, these being the T Tauri variables which are faint dwarf stars of late spectral type showing emission lines in their spectra and usually associated with small wisps of reflection nebulosity.

There is now ample evidence to suggest that all of these stars are very young, pre-main sequence objects still connected with their parent clouds. In the case of the T Tauri stars, at least, it is possible that some of them may possess protoplanetary systems.

All of these variables are found immersed in the large gas and dust clouds, in H I regions and, sometimes, very close to H II regions. It is therefore not surprising to find them in stellar associations, especially in the constellations of Orion, Auriga and Taurus in the northern hemisphere and in Carina, Corona Australis and Monoceros south of the celestial equator.

In many cases we find variables of all three groups in the same region and consequently, with the fainter members, it is not easy to determine into which class they should be placed. Much of this difficulty arises from two causes. Lack of sufficient observations to define the light curve accurately and over long periods; and a deficiency of spectroscopic data. Both of these combine to make the task of classifying the fainter variables extremely difficult.

One further problem has been (and sometimes still is) the indiscriminate use of the terms RW Aurigae, T Orionis and T Tauri variables for these stars. There is, unfortunately, no completely satisfactory scheme for the classification of the nebular variables and therefore the author has had to resort to a somewhat arbitrary division based mainly upon the characteristics of the light curves and, to a lesser extent, upon their spectra. While it is recognized that this is still not the ideal solution to the problem, it does have the advantage of basing such a differentiation upon the most readily obtained parameters of these stars.

J. S. G.

Acknowledgements

COURTESY of the Royal Astronomical Society: *Monthly Notices of the Royal Astronomical Society* and the authors named.

Figure 13 taken from *M.N.R.A.S.* **161**, 98 (Cohen); Fig. 16 taken from *M.N.R.A.S.* **161**, 234 (Moss); Fig. 17 taken from *M.N.R.A.S.* **145**, 278 (Larson); Fig. 18 taken from *M.N.R.A.S.* **157**, 130 (Larson); Fig. 19 taken from *M.N.R.A.S.* **157**, 134 (Larson); Fig. 25 taken from *M.N.R.A.S.* **161**, 103 (Cohen); Fig. 31 taken from *M.N.R.A.S.* **157**, 129 (Larson); Fig. 32 taken from *M.N.R.A.S.* **157**, 130 (Larson); Fig. 40 taken from *M.N.R.A.S.* **161**, 102 (Cohen); Fig. 41 taken from *M.N.R.A.S.* **161**, 100 (Cohen); Fig. 42 taken from *M.N.R.A.S.* **161**, 102 (Cohen); Fig. 48 taken from *M.N.R.A.S.* **157**, 127 (Larson); Fig. 49 taken from *M.N.R.A.S.* **157**, 128 (Larson); Fig. 50 taken from *M.N.R.A.S.* **157**, 128 (Larson); Fig. 63 taken from *M.N.R.A.S.* **161**, 283 (Robinson *et al.*); Fig. 64 taken from *M.N.R.A.S.* **161**, 285 (Robinson *et al.*); Fig. 66 taken from *M.N.R.A.S.* **145**, 301 (Larson); Fig. 67 taken from *M.N.R.A.S.* **145**, 304 (Larson) and Fig. 68 taken from *M.N.R.A.S.* **145**, 305 (Larson).

Courtesy of Professor L. Detre, Konkoly Observatory: *Non-Periodic Phenomena in Variable Stars*, IAU-Colloquium, Budapest (1969).

Figure 14 taken from page 91 (Seggewiss & Geyer); Fig. 15 taken from page 108 (Dibaj & Esipov); Fig. 23 taken from page 65 (Wenzel); Fig. 24 taken from page 66 (Wenzel); Fig. 26 taken from page 175 (Rosino); Fig. 27 taken from page 176 (Rosino); Fig. 28 taken from page 145 (Andrews); Fig. 29 taken from page 166 (Mirzoyan & Parsamian); Fig. 30 taken from page 108 (Dibaj & Esipov); Fig. 35 taken from page 69 (Wenzel); Fig. 36 taken from page 69 (Wenzel); Fig. 38 taken from page 72 (Wenzel); Fig. 43 taken from page 97 (Anderson & Kuhi); Fig. 44 taken from page 98 (Anderson & Kuhi); Fig. 45 taken from page 98 (Anderson & Kuhi); Fig. 46 taken from page 107 (Dibaj & Esipov); Fig. 47 taken from page 107 (Dibaj & Esipov); Fig. 54 taken from page 77 (Herbig); Fig. 58 taken from page 406 (Boyarchuk); Fig. 59 taken from page 408 (Boyarchuk).

Courtesy of the American Association of Variable Star Observers: *Q. Rep. A.A.V.S.O.* (Mrs. M. Mayall).

Figures 6, 7, 8, 9, 10 and 61.

It is also a pleasure to thank the following for their permission to quote extensively from their various papers, Dr. M. Cohen, Dr. R. B. Larson and Dr. D. L. Moss.

Introduction

History

The first variable star discoveries were made during the pre-telescopic era and were, quite naturally, accidental naked-eye observations and relatively few in number. Following the invention of the telescope an increasing number of variables were found and systematically observed, but it was not until the middle of the nineteenth century that the study of variable stars gained any real impetus.

The lack of accurate star charts was gradually overcome, culminating in the publication of the *Bonner Durchmusterung* and its two supplements. The introduction of photography, coupled with the blink comparator, further simplified the discovery of new variables. Indeed, so rapidly has this technique developed that the *General Catalogue of Variable Stars* (*GCVS*) listed 10,912 variables in its first edition (1948) and 20,437 in the third edition (1968). There is also a *Catalogue of Stars of Suspected Variability* (*CSSV*) containing 8134 objects.

Classification

The light variations of a variable star, as plotted in the form of a light curve, often enable a star to be assigned to a particular class merely by inspection. In borderline cases, the spectrum almost invariably provided the necessary additional information.

For convenience, we may divide variable stars into three broad groups.

The extrinsic variables. These are stars whose light variations are due more to a fortuitous set of external causes than to any intrinsic changes in the stars themselves. In this group we include the various classes of eclipsing binaries: the Algol, β Lyrae and W Ursae Majoris stars. In these binary systems the plane of the orbit lies in, or very close to, the line of sight as seen from Earth and consequently the component stars mutually eclipse each other during the course of their revolution about their common centre of gravity.

The idea that these geometrical variables should not be included among the true variable stars has recently been modified, if not abandoned altogether. There are clearly very complex physical changes taking place in the atmospheres of these stars and mass exchange between one component and the other, together with limb darkening and the well-known reflection effect, require intensive study.

The pulsating variables. These are mainly single stars in which various internal changes bring about pulsations and atmospheric shock waves that combine to produce the observed variations in brightness and spectral characteristics. Among these stars we find the RR Lyrae, Cepheid, long period, semi-regular and irregular variables. In contrast to those stars in the preceding group, these may also be termed "intrinsic variables" since their variability is not the result of a purely fortuitous set of external circumstances.

The eruptive variables. Here we include the dwarf novae, recurrent novae, novae, supernovae, flare stars (UV Ceti variables) and R Coronae Borealis variables. The light fluctuations

1

The Nebular Variables

in these stars are due to the ejection, sometimes explosive in nature, of matter from the stars. The dwarf novae and R Coronae Borealis variables, although showing little spectroscopic evidence for such ejection of mass (in the form of measured radial velocities), do either have light curves very similar to the novae and recurrent novae (dwarf novae), or show evidence for the condensation of solid material at some distance from the star (R Coronae Borealis variables).

Solely on the basis of their light curves, the nebular variables with which we shall be concerned are usually included among the irregular variables. As we shall see, however, their light fluctuations are due to both violent convective and chromospheric activity in the stars themselves and to external changes in their immediate environment.

All of these stars are intimately connected with dark or bright nebulosity which, itself, makes a major contribution to the light and spectroscopic characteristics. Since we have this combination of these two effects it is not surprising to find that the light curves of these stars show highly irregular and complex features.

A further distinguishing feature of most, if not all, of the genuine nebular variables (excluding such stars as R Aquarii and η Carinae) is that they are stars still in the pre-main sequence evolutionary stage, contracting out of the gas and dust clouds in which they are immersed. Another criterion of such variable contracting stars is their peculiar spectrum. At low dispersion, for example, the presence of the Hα line in emission has been employed during extensive surveys in the search for very young stars.

Nomenclature

GENERAL

By the middle of the last century, when the investigation of variable stars resulted in more and more being recognized and studied, it became apparent that there was a pressing need for some means of differentiating variable from non-variable stars. The method finally adopted was to designate variables with capital Roman letters beginning with R. Unfortunately, the letters from A to Q had already been employed for non-variable stars in the newly-charted southern constellations.

Accordingly, the first variable to be discovered in any particular constellation is given the letter R, followed by S, T to Z. The sequence then continues with doubled letters beginning with RR, RS . . . RZ; SS, ST . . . SZ and so on to ZZ. Where more variables are discovered, the sequence goes on with AA, AB . . . AZ; BB, BC . . . BZ until it finally ends with QZ. Where such doubled letters are used, the second letter is always later in the alphabet than the first.

This sequence provides a total of 334 combinations (the letter J being omitted to avoid confusion) and thereafter the variables are designated with the letter V and a running number. Examples are R Andromedae, RR Tauri, V1016 Cygni.

Until the variability has been independently confirmed and the above nomenclature universally adopted, a star suspected of being variable is allotted letters by the observatory detecting possible variability, e.g. BV . . . = Bamberg Variable; HV . . . = Harvard Variable; S . . . = Sonneberg Variable and VV . . . = Vatican Variable.

NEBULAR VARIABLES

The variables we shall be dealing with here are not easy to classify except in a rough and empirical manner. Representatives of the various types have been known for a long time,

2

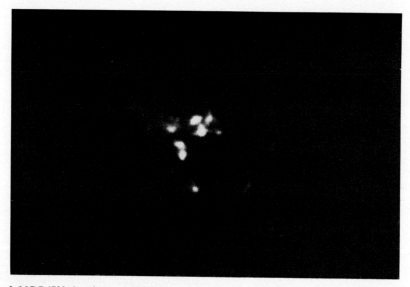

PLATE I. NGC 6729 showing variability between 1913 and 1916. (Helwan Observatory Photograph.)

but until comparatively recently they were concealed among the general groups of "Nebular Variables", "RW Aurigae" and "T Tauri" stars.

Some confusion has existed, and still does exist, in the classification of these variables. This is, perhaps, inevitable when we consider that we are dealing with stars having quite a large range of masses, luminosities and evolutionary ages. Quite a number of different classification schemes have been put forward, none of which is completely satisfactory.

That recommended by Commission 27 of the IAU is a compromise in many respects, classifying them as irregulars and according to their spectral type, kind of light variation and whether or not they are related to diffuse nebulae.

Here, these variables will be divided into four broad classes; RW Aurigae, T Orionis, T Tauri and Peculiar nebular objects. Classification according to the first three of these groups is based primarily upon the characteristics of the light curves and, to a lesser extent, upon the spectral type. Since all of these stars are immersed in nebulosity and the majority are found in the various T-associations (extensive groupings of faint nebular variables), it is inevitable that many of the fainter members have not yet been rigorously assigned to any particular class. A great deal of observational data on masses, luminosities, light variations and spectral types are required for many of these stars before they can be placed unambiguously in one or other of the above classes.

Among the peculiar nebular variables are included such stars as R Aquarii, FU Orionis and η Carinae which, although fulfilling the criterion of close association with nebulosity, clearly do not fall within any of the three other groups.

We must also take into consideration the numerous flare stars that have been discovered beyond the immediate neighbourhood of the Sun since these flare stars, in common with many T Tauri variables, are strongly concentrated towards the centre of the Orion Nebula and, almost beyond doubt, T-association members of Orion T-2 as shown by Kholopov (1959).

If, as appears likely from the work of Haro and Chavira (1965), such flare stars represent the evolutionary stage immediately following the T Tauri stage, then the connection between these flare stars and the genuine nebular variables of this type is obvious. In this context, it is perhaps significant that according to Mirzoyan and Parsamian (1969) none of the flare stars discovered in the small, compact T-association near NGC 7023 coincide with known nebular variables in this region. On the assumption, however, that all of these flare stars originated in the central region of this T-association, then their distribution does confirm Haro's hypothesis.

Techniques for Discovery and Observation

VISUAL OBSERVATION

Several of the brighter nebular variables, e.g. RW Aurigae, T Orionis, T and RR Tauri, have been visually observed for many decades by members of the various amateur associations. Visual estimates made by experienced observers are, in general, accurate to within $\pm 0^m.1$ and are therefore sufficiently precise to define the overall light curve provided that a sufficient number of observations are available and the various physiological problems associated with visual estimates are taken into account. The problem can become complicated at times by the fact that, where the nebular variables are concerned, the amplitude of variability is subject to certain long-term variations whereby, at times, the star may remain essentially constant in brightness. A case in point is T Tauri itself which for many years had an amplitude of $3^m.8$ in the visual but which now varies by only $0^m.4$.

The Nebular Variables

The large majority of the nebular variables are now detected, and followed, by photographic techniques. Apart from the obvious fact that a photograph contains the images of several hundred stars, provides a permanent record and reaches fainter magnitudes, the discovery of new variables is greatly facilitated by the use of the blink comparator.

Wide-angle objectives not only cover a large area of sky but can also reach down to $\sim 14^m$ as a general limit. For observation of fainter variables, larger instruments are necessary, although these have the restriction that they cover limited fields. The use of a suitable yellow filter and photographic emulsion provides photovisual magnitudes which are comparable to visual ones.

PHOTOELECTRIC OBSERVATIONS

Many of the nebular variables are covered by a series of photoelectric observations since, quite often, the light fluctuations of these stars have both small amplitudes and short periods. A typical photoelectric cell has a multiplication of approximately 1,000,000 and a corresponding accuracy which is much greater than that obtained by visual or photographic methods.

With the recent development of efficient detectors, the germanium bolometer cooled by liquid helium and photoconductive cells of silicon, indium antimonide or lead sulphide cooled by liquid nitrogen, it is now possible to carry out observations in the infrared. This is of particular importance in the case of the nebular variables which possess a high infrared flux. Corresponding to the UBV system of magnitudes we therefore have an extension into the infrared, the designations and wavelengths being as given in Table I.

TABLE I. INFRARED SYSTEM OF MAGNITUDES

Designation	Wavelength (μ)
I	0.90
J	1.25
K	2.20
L	3.60
M	5.00
N	10.80
Q	20.00

The units of flux density used in infrared work are in $W\ m^{-2}\ Hz^{-1}$.

Far infrared observations are made with gallium-doped, germanium bolometers cooled to liquid helium temperatures. Recent observations in this region have been made with balloon-borne instruments, e.g. those by Furniss et al. (1972) and Emerson and his colleagues (1973).

It is also possible to measure ultraviolet energy fluxes for variable stars, comparing these with model atmospheres for main sequence stars. Observations are made at a height of 80–100 miles since the terrestrial atmosphere absorbs much of the ultraviolet radiation. Photomultipliers coated with sodium salicylate are normally used in this work and measurements are generally made at wavelengths of 2800, 2500, 2100 and 1376 Å.

In the case of ultraviolet energy fluxes, the units used are $erg\ cm^{-2} sec^{-1}\ Hz^{-1}$.

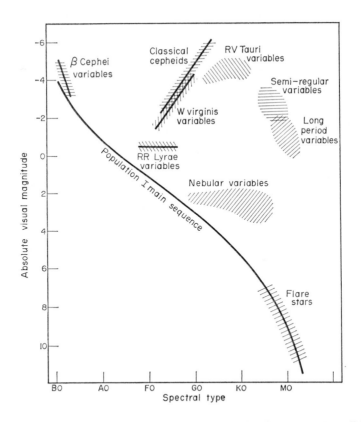

Fig. 1. Positions of various classes of intrinsic variables on the Hertzsprung–Russell diagram.

RADIOMETRIC OBSERVATIONS

The radio emission from celestial objects is collected by a dipole situated at the focus of a parabolic mirror. The high-frequency waves are then amplified and rectified and usually recorded as either intensity or flux density by means of a suitable recording system.

The positions of the various emission lines in the radio region are given either as wavelength (cm) or frequency (MHz or GHz). The units of radio density are given in W m^{-2} Hz^{-1}.

The emissions from various interstellar molecules are found in the radio region of the spectrum, e.g. hydroxyl (OH) at 1612, 1665, 1667, 1720 MHz; ammonia (NH$_3$) at 23.6 GHz; and formaldehyde (H·CHO) at 4830 MHz. The detection of ammonia has been reported by Cheung *et al.* (1969), and that of formaldehyde by Snyder *et al.* (1969) and Palmer and his co-workers (1969).

SPECTROSCOPIC OBSERVATION

As we shall see, one characteristic of most of the nebular variables is the presence of Hα in emission. Low-dispersion spectrograms, obtained by using an objective prism, are often used in surveys for detecting these variables. High-dispersion spectrograms of relatively bright objects are usually obtained by means of a slit spectrograph. The latter are of particular importance in the study of these stars since their spectra are unusually rich in

5

The Nebular Variables

peculiarities due, not only to the stars themselves, but also to their extensive atmospheres and the surrounding circumstellar clouds with which they are intimately associated.

THE HERTZSPRUNG–RUSSELL DIAGRAM

The positions of the nebular variables on the Hertzsprung–Russell diagram, particularly their relation to the main sequence, are important since the large majority are still in the pre-main sequence stage of contraction. The location of the T Tauri variables, for example, in relation to other types of intrinsically variable stars, is shown in Fig. 1.

HIGH-SPEED PHOTOMETRY/SPECTROSCOPY

Nather and Warner (1971a, 1971b) have demonstrated how modern high-speed electronic techniques can extend the usual UBV photometry of various classes of variable stars, in particular certain of the symbiotic variables. Very rapid light fluctuations with amplitudes of only 0.01 magnitude can be readily measured by this technique.

References

CHEUNG, A. C., RANK, D. M., TOWNES, C. H., KNOWLES, S. H. and SULLIVAN, W. T. (1969) *Astrophys. J. Lett.* **157**, L13.

EMERSON, J. P., JENNINGS, R. E. and MOORWOOD, A. F. M. (1973) *Nature Phys. Sci.* **241**, 108.

FURNISS, I., JENNINGS, R. E. and MOORWOOD, A. F. M. (1972) *Ibid.* **236**, 6.

HARO, G. and CHAVIRA, E. (1965) *Vistas in Astronomy* **8**, 89.

KHOLOPOV, P. N. (1959) *Soc. astr. A.J.* **3**, 291.

MIRZOYAN, L. V. and PARSAMIAN, E. S. (1969) *Non-Periodic Phenomena in Variable Stars*, p. 165. Reidel, Dordrecht.

NATHER, R. E. and WARNER, B. (1971a) *Mon. Not. Roy. astr. Soc.* **152**, 209.

NATHER, R. E. and WARNER, B. (1971b) *Ibid.* **152**, 219.

PALMER, P., ZUCKERMAN, B., BUHL, D. and SNYDER, L. E. (1969) *Astrophys. J. Lett.* **156**, L47.

SNYDER, L. E. and BUHL, D. (1969) *Astrophys. J. Lett.* **155**, L65.

CHAPTER 1

General information on Nebular Variables

History

As the study of variable stars progressed during the nineteenth century, certain variables were discovered whose light curves showed no trace of periodicity. The early investigators tended to discuss each star individually rather than to classify them by type and, since most of the variables then known exhibited some form of regularity in their light fluctuations, it was expected that all of them should do so. As a result, observers attempted to deduce periods for each new variable and in many cases this proved difficult, if not impossible.

John Herschel considered that totally irregular variation existed in certain stars and attributed this to the presence of interstellar clouds of gas and dust. Random changes in the opacity of this material, lying between us and the star in question, would produce similar random fluctuations in stellar brightness. However, the majority of observers of the mid-nineteenth century worked on the principle that for every variable star a period existed, although it might be concealed under various superimposed irregularities and hence difficult to elucidate.

About this time, as more variables were discovered and comprehensively observed, many were found to exhibit almost identical modes of light variation and could thus be assigned to a number of reasonably well-defined classes. Irregular variables were characterized as those for which no prediction as to their epochs of maximum or minimum brightness could be made with any degree of accuracy. It is certainly unfortunate that this fact alone contributed to the almost complete cessation of observational work on these particular stars for a long time. The great majority of workers concentrated their efforts on those variables for which some degree of periodicity could be demonstrated, undoubtedly because of the greater accuracy and ease with which observations and predictions of these stars could be made.

Argelander (1848) was able to show that irregularities are a common feature of such long period variables as o Ceti which do exhibit well-pronounced periodicities and considered it only natural to search for periods in the cases of those variables having apparently irregular light variation.

The possibility that there exist stars which do not possess any period at all was seriously considered by Schmidt (1857), a contemporary of Argelander, and from this time onward the idea that a class of totally irregular variables exists became more prominent. Pickering (1881) put forward a scheme for the classification of variable stars which included those with non-periodic light variation. Normally, these were found to have only small amplitudes of between one and three magnitudes. A similar group was included in the more recent scheme put forward by Lundmark (1935).

The Nebular Variables

Classification

As will be seen from the light curves given throughout the following chapters, the nebular variables belong to the class of irregular variables. It will also become clear that they include a wide range of types which, in general, have only three features in common.

(a) They are mostly very young stars, still contracting gravitationally, which have not yet reached the zero age main sequence and whose spectra, at low dispersion, normally show the Hα line in emission.

(b) With the exception of some early spectral type T Orionis stars and peculiar nebular objects such as η Carinae, they possess low intrinsic luminosities.

(c) They are all associated, in some way or other, with either dark or bright nebulosity.

In this last respect it may be said that they represent those stars envisaged by John Herschel in which the light fluctuations are due, in part at least, to the presence of this nebulosity.

Stellar Associations

In the study of the nebular variables, the concept of stellar associations has become increasingly important since one of the chief characteristics of the stellar associations is that the large majority of the stars forming any one association have similar physical parameters. In particular, they are all of approximately the same age.

The stellar associations may be assigned to one of the following three classes: the O-associations which contain large numbers of hot giants, stars of spectral classes O and B, with relatively small numbers of other objects; the T-associations which contain mainly relatively cool dwarf stars, the majority of which are irregularly variable in brightness and form the greater proportion of the nebular variables, and the R-associations which contain young, massive variables that are probably the massive counterparts of the T Tauri stars.

Since they are made up of numerous intrinsically bright, giant stars, the O-associations may be observed at great distances. Many have already been discovered in the Greater and Lesser Magellanic Clouds and they are also obvious in the spiral galaxy M33 and the irregular galaxy IC 1613.

The region around the Trapezium in Orion also contains an O-association, another being Perseus I which consists of the double cluster h and χ Persei. Close groups of stars are found within each O-association in the form of trapezia, chains and small open clusters and these have been termed the nuclei of the association. The Orion O-association also contains nebulosities which radiate in the visible and ultraviolet regions due to the effect of the hot stars which are embedded within them.

While it is true that many of the O type stars found in the O-associations are variable in brightness, their amplitudes seldom exceed a few hundreds of a magnitude.

The T-associations generally have a close connection with dark nebulae and contain RW Aurigae, T Orionis and T Tauri variables. The question of whether the T-associations also contain non-variable dwarf stars is clearly an important one, but unfortunately it is not an easy question to answer since dwarf stars of constant brightness belonging to T-associations cannot readily be distinguished from the multitude of similar dwarfs belonging to the general stellar population.

The dimensions of the T-associations have been found to lie between a few parsecs and more than 100 pc. In some cases their stellar density is quite high, while in others they are more rarefied and somewhat less dense than the surrounding star field.

The R-associations are found in reflection nebulae and contain Be and Ae type stars which appear to be still in a rotationally unstable pre-main sequence evolutionary stage. There appears to be little doubt that these associations are also young stellar aggregates with ages of $\sim 10^6$ years, i.e. comparable with both the O- and T-associations.

THE COMPACT GALAXY II ZWICKY 40

Here we may mention the very compact galaxy II Zwicky 40 which has been studied by Jaffe (1972) and Gottesman (1972). The former has shown that there is a detectable radio emission from this object at 1415 MHz. The measured flux density combined with H I measurements indicate a diameter of ~ 1 kpc. The kinetic energy of the galaxy seems to be primarily non-rotational and the system is probably less than 10^7 years old.

Gottesman has found that the neutral hydrogen follows a core–halo distribution. The lower mass limit obtained from the dynamics of the system is $9.7 \times 10^8 \, M_\odot$ for a distance of 12.8 Mpc. The low metallicity and extreme blue colour of the galaxy, coupled with the large dimensions of the H I region, imply that only recently has star formation begun in very restricted regions. A very bright optical object lies within the broader H I core.

Unquestionably the nebular variables form a very heterogeneous group. As noted by Herbig (1962), this is particularly so for the RW Aurigae stars which embrace a wide range of spectral types and luminosities. In essence, however, the light curves of these stars are characterized by fairly rapid and non-periodic fluctuations. Some approximately periodic light phenomena have been found, chiefly by spectral analysis of their light curves.

The RW Aurigae and T Orionis variables have quite large amplitudes of the order of 4^m although, even here, we usually find that for much of the time they vary by only one or two magnitudes. The T Tauri variables, on the other hand, generally have somewhat smaller amplitudes, particularly in the visible. Their infrared variations, however, are often appreciably larger than their visual amplitudes.

Since one of the main criteria for assigning a star to the class of nebular variables is that it should be intimately connected with some form of nebulosity, it is inevitable that we should also find some stars which may be regarded as peculiar in that they do not fit into a simple scheme of classification.

For example, we have such stars as R Aquarii which is normally classed as a long period variable of the Mira type, similar in most respects to the prototype star o Ceti. The spectrum, however, and certain peculiarities discovered in its light curve, reveal that it is a much more complex object consisting of a red giant; a hot, white subdwarf and a nebulous envelope.

The symbiotic variables are also included in this group by virtue of the fact that they possess some of the characteristics of stars such as R Aquarii. The light curves show irregular variations with the increase in brightness being steeper than the decline. In their spectra, both permitted and forbidden emission lines develop during the decline to minimum, these being of progressively higher excitation and ionization. We find, for example, the lines of H, He II, [N III], [O III] and even more highly ionized atoms. This development of the emission lines is characteristic of ejection of matter into a surrounding envelope.

In addition to the genuine symbiotic variables, we may include the recurrent novae such as T Coronae Borealis and RS Ophiuchi. The original model for the recurrent novae was that of a single star from which a relatively small amount of mass is ejected during a typical outburst.

High-dispersion spectrograms have now shown that the evidence is more compatible with their being spectroscopic binaries. The radial velocity curves, for example, have

The Nebular Variables

periods of several hundreds of days. This model has now been confirmed from the doubling of the lines which belong to a red giant or supergiant component of spectral type Me III and a blue star, the emission line of $H\beta$ being used to derive the velocity curve.

Other peculiar objects that also come under the same general heading of the nebular variables are the Herbig–Haro objects which appear to be strange semi-stellar emission nebulae first detected by Herbig (1948, 1951) and Haro (1950, 1952). All of these objects that have so far been discovered lie in regions that are heavily obscured by nebulosity and which contain large numbers of T Tauri variables.

We shall also consider the infrared objects that are being discovered in increasing numbers; these being either very young stars gravitationally contracting out of gas and dust clouds, or early main sequence stars which are still surrounded by a dust shell that absorbs the underlying stellar radiation and re-emits it in the infrared.

Because we are dealing, in the main, with stars having a large range of masses and evolutionary ages and exhibiting different forms of variability, this does not permit the use of a simple scheme of classification. Several such schemes have been suggested based upon either photometric observations or differences in spectra. The most recent, recommended by Commission 27 of the IAU in 1964 for very young stars, classifies them according to the following three principles:

(a) The spectral type, whether this is early, intermediate to late, or similar to T Tauri.

(b) The variation in brightness which may be rapid, slow or exhibit flare-like activity.

(c) Whether or not the star is physically associated with diffuse dark or bright nebulosity.

This scheme of classification, however, has certain disadvantages as pointed out by Wenzel (1969). Primarily it does not fully differentiate between the obviously different light curves of variables such as RW Aurigae and T Orionis. Both of these stars would be classed as "Ins" type variables, especially if their spectra were unknown. In other words, they are irregular variables found in, or near, diffuse nebulae and varying in brightness by approximately $1^m.0$ in a period of several hours or days.

An earlier classification was made by Parenago (1950) based upon the characteristics of the light curve and the logarithm of the time scale from one maximum brightness to the next. Parenago's classes are as follows:

Class I   Variable more frequently near maximum than at minimum.
Class II   Variable mostly around the mean brightness.
Class III   Variable more frequently at minimum than maximum.
Class IV   Variable found at any brightness within its range.

For those cases in which the variable has an unresolved companion, which often results in the observed amplitude being smaller than the true amplitude of the star itself, the classification is suffixed by the letter c. In the case of RW Aurigae, for example, we find evidence for light cycles of approximately 5 days and also for the presence of an unresolved companion. On the above scheme, therefore, this variable is Class II $-0.5c$.

Here, we shall classify these very young stars, now numbering between three and four hundred, as RW Aurigae, T Orionis and T Tauri variables, this scheme being similar to that advocated by Wenzel (1969).

Although the fluctuations are highly irregular, when we examine the light curves in detail it is apparent that there are certain broad similarities which, when taken in conjunction with their spectra, enable these stars to be placed in the above classes with the following characteristics.

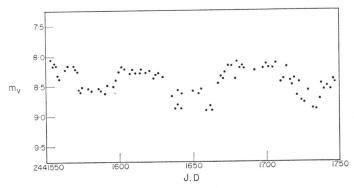

Fɪɢ. 2. Light curve of HH Aurigae.

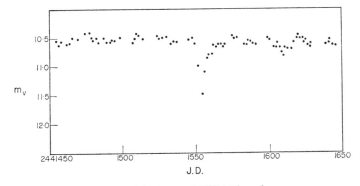

Fɪɢ. 3. Light curve of WW Vulpeculae.

The RW Aurigae Variables

The light curves of these stars show that they spend approximately the same time at maximum as at minimum. In certain cases, too, there is a distinct tendency to remain for relatively long periods at some brightness intermediate between the two extremes. This type of variation is clearly shown in the light curve of HH Aurigae (Fig. 2) which is typical of the RW Aurigae class.

The spectra are usually of types G and K, with or without emission lines. A small number of these stars have earlier spectra (V852 Ophiuchi, A2e; RR Tauri A2 II–III). Most, if not all, of these variables are dwarfs.

The T Orionis Variables

These are stars which remain at, or near, maximum brightness for most of the time but which are subject to abrupt and non-periodic minima. In one or two instances, this latter feature is so marked that the light curve resembles that of an eclipsing binary and in one case, that of B0 Cephei, the star has been erroneously assigned to the Algol class before its true nature was recognized. Slow and short fluctuations are a common feature near maximum.

A large number of these variables are found within the Orion complex and other bright nebulae. Figure 3 illustrates the light curve of WW Vulpeculae showing both types of light variation.

11

The Nebular Variables

The spectra of these stars cover a wider range than those of the preceding group, from types B to K and, as before, we find representatives both with and without emission lines in their spectra. Unlike the RW Aurigae variables, many of these stars are giants with surface temperatures in the range 7500–10,000°K. A few members, e.g. TY Coronae Australis, KX and LP Orionis, have an appreciably higher surface temperature of the order of 25,000°K.

THE T TAURI VARIABLES

An examination of the light curves of many of these stars reveals the presence of a rapid cycle (period between 1 and 10 days) superimposed upon a slow component (period 90–200 days). Although they can have amplitudes of up to $4^m.0$, there are often long periods, sometimes of several years, during which the visual brightness varies by much less than a magnitude. Over the past two decades, the prototype star T Tauri has varied only between $10^m.1$ and $10^m.6$, whereas in the past the amplitude amounted to almost four magnitudes (Fig. 4).

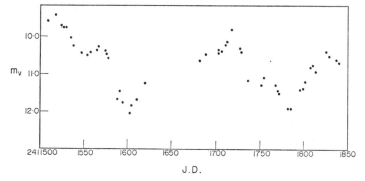

FIG. 4. Light curve of T Tauri showing presence of large amplitude.

The spectra of the T Tauri variables are predominantly later than those of the two preceding groups and all appear to show emission lines. All are dwarf stars, mainly associated with dark and variable nebulae and are possibly the youngest of the three classes. There now seems little doubt that the light variations of the T Tauri stars and their associated nebulae are intimately connected. This is to be expected since, on average, the physical association of star and surrounding nebula will decrease with increasing stellar age. In certain cases there is also fairly strong observational evidence that these variables are surrounded by a ring of dust that may be a planetary system in the making.

One important feature of the T Tauri stars is that they are distinctly brighter than main sequence stars of the same spectral type, further confirmatory evidence that they are still in the gravitationally contracting stage.

RY Tauri, for example, which is now a dwarf star with a spectral type of dG0e, will reach the main sequence in about 1×10^7 years, becoming hotter and smaller in the process until it has a spectrum approximating type A7. If the angular momentum is conserved throughout this period, this will result in a doubling of its rotational velocity to about 100 km/sec. This is the same order of rotational velocity as found for stars of spectral type A7 on the main sequence.

12

Those T Tauri variables with later spectra of types K and M such as AA Tauri (spectral
type dM0e), will evolve into dG0 stars by the time they reach the main sequence and
theoretical considerations indicate that in their case they will become more slowly rotating
bodies with equatorial velocities very similar to that of the Sun, again in good agreement
with observation.

Spectral Analysis of Light Curves

The light variations of the nebular variables as a group are often apparently completely
irregular with many rapid fluctuations superimposed upon longer-term variations. Consequently it is very difficult to determine any periodicities which may be present but obscured
by noise. Since it is virtually impossible to compute theoretical light curves for these variables,
the use of power spectrum analysis of the empirical data is a powerful tool for revealing the
presence of peaks above the general noise level and providing an exact value for any periodicity and its amplitude.

Plagemann (1969) has prepared power spectra from the light curves of 130 nebular
variables given in the table by Herbig (1962). Since none of these variables shows strictly
periodic variation in the various, superimposed light changes, they represent a mixed time
series and the method of harmonic analysis cannot therefore be used. In fact, as is well
known from statistical analysis, harmonic analysis of noise or random numbers produces a
highly spiked spectrum.

Methods for accurately estimating time series such as those represented by the irregular
light curves of the nebular variables have been developed by Blackman and Tukey (1958)
and Weiner (1967).

The method given by Blackman and Tukey is applicable to the study of short-term
fluctuations in these stars. However, it can be easily shown that it cannot readily be applied
to the examination of the longer-term changes since the sampling rate is drastically dis-
turbed by inevitable gaps in the observations (due to the day–night variation) unless a
variable is constantly watched on a 24-hour basis by observatories all over the world. This
is obviously not a practical proposition for a large number of these stars.

Jenkins (1965) has suggested an empirical method which reduces these difficulties to a
certain extent.

In addition to deriving any changes in the period of the light variations and also their
amplitude, the information gained from such power spectra can be compared with other
properties of the nebular variables, e.g. their spectral types, peculiarities in their absorption
and emission spectra, stellar age and mass when compared to a theoretical Hertzsprung–
Russell diagram.

References

ARGELANDER, F. A. W. (1848) *A.N.* **26**, No. 624, 369.
BLACKMAN, R. B. and TUKEY, J. W. (1958) *The Measurement of Power Spectra*, Dover Publications.
GOTTESMAN, S. T. (1972) *Astrophys. Lett.* **12**, 63.
HARO, G. (1950) *Astron. J.* **55**, 72.
HARO, G. (1952) *Astrophys. J.* **115**, 572.
HERBIG, G. H. (1948) Thesis, Univ. of California.
HERBIG, G. H. (1951) *Astrophys. J.* **113**, 697.
HERBIG, G. H. (1962) *Adv. Astr. Astrophys.* **1**, 47.

The Nebular Variables

JAFFE, W. J. (1972) *Astron. and Astrophys.* **20**, 461.
JENKINS, G. M. (1965) *Applied Statistics* **14**, 205.
LUNDMARK, K. (1935) *Lund. Medd. Ser. II* No. 74.
PARENAGO, P. P. (1950) *Peremennye Zvezdy*, **7**, 169.
PICKERING, E. C. (1881) *Proc. Am. Acad.* No. 16.
PLAGEMANN, S. (1969) *Non-Periodic Phenomena in Variable Stars*, p. 21. Reidel, Dordrecht.
SCHMIDT, J. (1857) *A.N.* No. 37, 164.
SCHMIDT, J. F. (1957) *A.N.* No. 1060, 61.
WEINER, N. (1967) *Time Series*, Dover Publications.
WENZEL, W. (1969) *Non-Periodic Phenomena in Variable Stars*, p. 61. Reidel, Dordrecht.

PART I
RW Aurigae Variables

CHAPTER 2

Light variations of RW Aurigae stars

In the previous chapter we saw how the apparently irregular variations found in the light curves of the nebular variables are, nevertheless, often sufficiently characteristic to enable astronomers to subdivide these stars into at least three distinct groups. The prototype stars of these groups are RW Aurigae, T Orionis and T Tauri.

Here we shall consider the light variations of the first group, both as regards their long-term variations as shown by visual, photographic or photometric observation; and the more rapid fluctuations in brightness that have been measured photoelectrically. We may also compare these with the closely allied T Orionis variables which, together with the T Tauri stars, are often found together in the large T-associations.

As may be seen from the data given in Table II, the amplitudes of these stars normally lie between 2^m and 4^m and, from the light curve of RW Aurigae (Fig. 5), it will be noted that the fluctuations are all extremely complex.

The overall light curve of a variable such as RW Aurigae appears to be completely irregular with the variations being totally unpredictable. We must bear in mind, however, that a light curve of this nature, plotted solely from visual estimates (or in the case of the fainter members, from photographic determinations), does not show the more rapid, short-term variations which are provided by photoelectric observations. These are naturally lost in any curve that depends upon estimates made at least a day apart.

It is only when we have a fairly long series of accurate photoelectric determinations, made at frequent intervals, that any quasi-periodic fluctuations show themselves. The most comprehensive light curve for RW Aurigae, prepared from numerous visual and photographic estimates by various authors, is that given by Kholopov (1962).

Although it is difficult, and perhaps unwise, to generalize for such irregular changes in brightness, it appears that the light curves of the RW Aurigae stars resemble those of the T Orionis variables more so than the T Tauri stars. This may be demonstrated by a comparison of the curves for RW Aurigae, R Monocerotis and CO Orionis with the light curves of typical T Orionis variables given in Figs. 6–9. It is also noticeable that the spectra of the RW Aurigae and T Orionis variables have a similar spread compared with the much narrower range in spectral types of the T Tauri stars. We also find that the latter all show typical dwarf characteristics in their spectra and invariably exhibit emission lines.

Visual Light Curves

By far the longest and most comprehensive series of observations of the brighter members of this group are visual estimates made by the various amateur organizations, e.g. the American Association of Variable Star Observers and the Variable Star Section of the Astronomical Association of New Zealand.

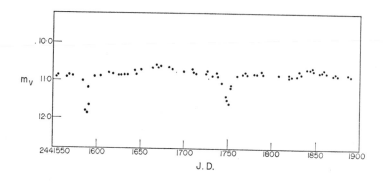

FIG. 5. Light curve of RW Aurigae.

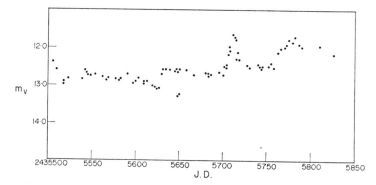

FIG. 6. Light curve of R Coronae Australis. (After Mayall.)

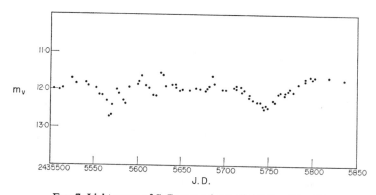

FIG. 7. Light curve of S Coronae Australis. (After Mayall.)

Owing to the physiological variations between one observer and another there is an inevitable scatter of points on these curves. This allows us only to draw mean curves which, in conjunction with the interval between observations, mask any small and rapid fluctuations.

If any generalities can be drawn from the available data, these are as follows: the stars are very seldom observed at the extremes of their light ranges; they remain for long periods

TABLE II. RW AURIGAE VARIABLES

| Star | Magnitude | | Spectrum |
	Maximum	Minimum	
RW Aur	9.0*	12.0	dG5e
UY Aur	11.6	14.0	dG5
HH Aur	8.9	10.3	G5
DI Car†	11.1	13.0	Pec: dK ?
ES Car	12.8*	14.5	B7 III
BE Cas	13.5	14.4	—
DR Cep	12.0	14.1	—
T Cha	10.0	13.2	dG5
BZ Del	13.6	14.4	—
BR Her	13.0	14.0	—
PQ Her	12.5	13.2	—
QT Her	12.4	13.6	—
RU Lup	9.3	13.2	dG5e
RY Lup	9.9	13.0	—
RX Lyn	10.3	12.7	—
R Mon	10.0*	14.2	G5
V852 Oph	14.5	15.0	A2e
V853 Oph	12.5	14.0	dKe
CO Ori	10.0	13.0	—
V426 Ori	12.2	13.8	dK0
SY Phe	9.5*	10.2	F8
SZ Phe	9.5*	10.7	K4
TT Phe	10.1*	11.3	dK?
RR Tau	10.2*	14.2	A2 II–III
BS Vel	13.2	14.5	dK ?

*Visual magnitudes, all others being photographic.
†Probably a Cepheid variable (P ~ 29^d).

at some intermediate brightness usually slightly brighter than the median magnitude, with only small fluctuations in brightness ($\sim 0^m.4$); they are subject to rather abrupt minima reminiscent of the T Orionis variables (amplitude $1^m.2$, duration 2–5^d).

RW AURIGAE

The prototype star is also one of the brightest of these variables although fewer observations appear to have been made of it than several other, fainter, members of this class. It is well established, however, that for long periods this variable exhibits cycles in its light variations (amplitude $0^m.5$–$1^m.0$, period $\sim 5^d$) which are quasi-periodic in character. Superimposed upon these fluctuations are often longer cycles which are, in general, much less well defined (Fig. 5). The irregular variations in brightness (with reasonably large amplitudes) are somewhat more pronounced in this star than in the majority of these variables.

R MONOCEROTIS

The extreme range of this variable is $4^m.2$ but an analysis of the light curve over the past three decades shows that for approximately 80 per cent of this period, the star varied only

The Nebular Variables

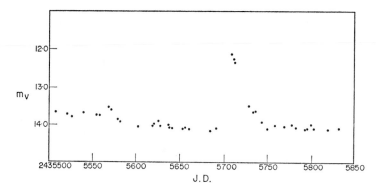

FIG. 8. Light curve of T Coronae Australis. (After Mayall.)

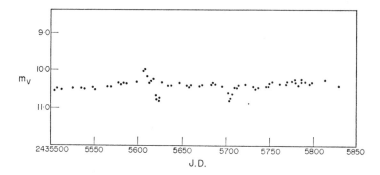

FIG. 9. Light curve of TY Coronae Australis. (After Mayall.)

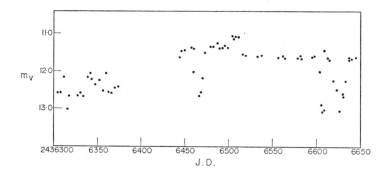

FIG. 10. Light curve of R Monocerotis. (After Mayall.)

between $11^m.1$ and $11^m.7$. Such behaviour appears to be fairly typical of the RW Aurigae stars.

Outside these periods of essentially constant brightness, the visual behaviour consists of non-periodic fadings to $\sim 13^m.0$ although isolated brightenings to $\sim 10^m.2$, similar to the outbursts of RW Aurigae, have been reported by Beidler (1959), Wend (1960) and Morgan (1962).

The light curve given in Fig. 10 is typical of the general behaviour of this star.

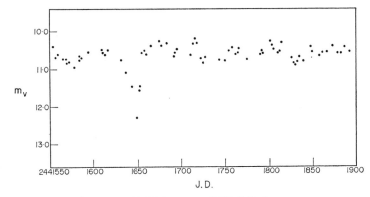

FIG. 11. Light curve of CO Orionis.

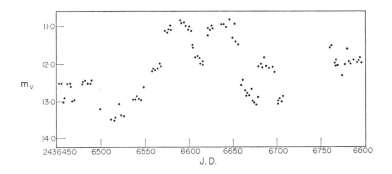

FIG. 12. Light curve of RR Tauri.

CO ORIONIS

The visual light variations of this variable (Fig. 11) are similar to those of R Monocerotis. From an analysis of the light curve, there appears to be little, if any, evidence for the comparatively short-period, quasi-periodic fluctuations which are shown by the prototype star.

Unfortunately, this variable has not been as assiduously observed as either of the two preceding variables and it is difficult to draw any statistical conclusions from the available light curve.

RR TAURI

While the majority of the RW Aurigae have spectra of intermediate type, RR Tauri is one of the comparatively few members (cf. ES Carinae and V852 Ophiuchi) with an early type spectrum. The visual light curve is fairly typical of this class. Periods of almost constant brightness are interspersed by relatively long minima (amplitude $\sim 1^m.7$).

The star has seldom been observed to flare to anything like its extreme brightness of $10^m.2$.

DI CARINAE

This star has been variously classified in the past. Hoffmeister (1957) formed the opinion that it belonged to the CN Orionis variables which form a subgroup of the U Geminorum

21

The Nebular Variables

(dwarf novae) variables. Kukarkin *et al.* (1958) classified it as an RW Aurigae variable. Recent photoelectric data obtained by Seggewiss (1970) led to the conclusion that the star is a classical Cepheid with $P = 29^d.210$.

The brightness variation of DI Carinae is larger than $0^m.7$ in V, $0^m.3$ in B–V and $0^m.34$ in U–B. There seems little doubt that the light curve more closely resembles that of a Cepheid than a typical RW Aurigae variable.

The spectrum has been examined by Seggewiss and Geyer (1969) who note three important features.

(a) There is a strong absorption feature that merges into the G-band from the longward side. It appears possible that this may be $H\gamma$ although if this is so, it seems too intense for the spectral type as estimated from the H and K lines of Ca I.

(b) From the undiminished intensity of the continuum in the region of 4125 Å, it appears that the line of Sr II is absent.

(c) Between $H\beta$ and $H\gamma$ the continuum is exceptionally weak. From this, it is concluded that the star may be a K type dwarf, quite possibly composite.

Photoelectric Light Curves

Although less extensive, these are, of course, much more accurate than light curves based upon purely visual estimates. They also provide data on the more rapid fluctuations with very small amplitudes.

RW Aurigae

The photoelectric variations of this star have been described by Wenzel (1966) and are much more complex than the visual light changes described above. The following components appear to be present:

(a) Long waves of several hours duration which have an amplitude of $\sim 0^m.4$.

(b) Minor flare-like outbursts, amplitudes $\sim 0^m.1$, which last for between 1 and 2 hours. These eruptions are all symmetrical in character.

(c) Unsymmetrical flares with amplitudes slightly larger than those of the symmetrical type, these possibly originating in the M type companion.

(d) Very small fluctuations with amplitudes between $0^m.04$ and $0^m.15$ which appear to be due to changes in the intensity of the emission lines.

(e) Quasi-periodic variations, amplitudes between $0^m.5$ and $1^m.0$ which have a cycle length of $\sim 3^d$. These changes in brightness have also been noted in the visual light curve.

Both the colour-luminosity and the two-colour diagrams for this star, constructed by Wenzel (1966), show a large intrinsic scatter. This is undoubtedly due to variations in the emission line intensity and the very abnormal continua of this star.

Southern RW Aurigae Variables

Photoelectric observations in the UBV system have been carried out by Seggewiss and Geyer (1969) for SY, SZ, TT Phoenicis, DI and ES Carinae and BS Velorum using the 60-inch Rockefeller reflector of the Boyden Observatory. Similar observations made earlier by de Kort (1941) and Hoffmeister (1958) showed that the three variables in Phoenix and also BS Velorum exhibit only small variations in magnitude and colour; results which were confirmed by the more recent estimates. No periodicity at all is apparent in the photoelectric light curves for these four variables.

DI Carinae clearly shows the more regular light variations typical of a Cepheid variable. In the B–V/U–B diagram, the star moves along a straight line above, and parallel to, the main sequence throughout its light variations. In both the V/B–V and V/U–B diagrams, it describes loops. This behaviour is somewhat similar to that of the Cepheids η Aquilae and BB Sagittarii as shown by Nikolov and Kunchev (1969).

The photoelectric observations in V for ES Carinae show an amplitude of $0^m.4$ with completely irregular variations.

Infrared Observations of RW Aurigae Variables

The brightness of R Monocerotis at wavelengths as far into the infrared as 20μ was discovered by Low and Smith (1966). Similar observations of the faint star LkHα-101 by Cohen and Woolf (1971) have confirmed the earlier suggestion by Mendoza (1966, 1968) that the major proportion of the radiation from these very young stars lies in the infrared beyond 2μ.

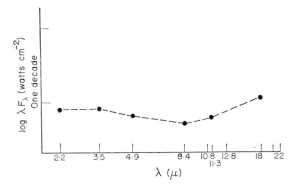

FIG. 13. Infrared energy distribution for CO Orionis. (After Cohen.)

Infrared observations of a number of nebular variables, including several of the RW Aurigae type, have been made by Cohen (1973a, 1973b). These variables include RW Aurigae, UY Aurigae, HH Aurigae, R Monocerotis, RR Tauri and CO Orionis. The infrared energy distribution curve for the latter star is shown in Fig. 13.

All of these variables exhibit very similar variations over the range from 2.2 to 11.0μ. The energy distribution curves are extremely flat and this characteristic is continued for CO Orionis out to 18μ.

Cohen (1973a) has pointed out that these curves are completely unlike those arising from either free–free emission or the radiation from a black body at a single temperature. Almost certainly, these observations provide further confirmatory evidence that, in these young gravitationally-contracting stars, there is some form of solid material surrounding the RW Aurigae variables which absorbs much of the stellar radiation from the central star and then re-radiates it by a thermal process in the infrared.

References

BEIDLER, H. B. (1959) *Q. Report A.A.V.S.O.* No. 23, 42.
COHEN, M. and WOOLF, N. J. (1971) *Astrophys. J.* **169**, 543.
COHEN, M. (1973a) *Mon. Not. Roy. astr. Soc.* **161**, 97.
COHEN, M. (1973b) *Ibid.* **161**, 105.

The Nebular Variables

DE KORT, J. (1941) *Bull. astr. Inst. Netherlands* **9**, 245.

HOFFMEISTER, C. (1957) Private communication in SCHNELLER, H., *Geschichte und Literatur des Lichtwechsels der veränderl. Sterne* **4**, Berlin.

HOFFMEISTER, C. (1958) *Veröff. Sonneberg* **3**, 348.

KHOLOPOV, P. N. (1962) *Peremennye Zvezdy* **10**, No. 6.

KUKARKIN, B. V. *et al.* (1958) *General Catalogue of Variable Stars*, MOSCOW.

LOW, F. J. and SMITH, B. J. (1966) *Nature* **212**, 675.

MENDOZA, E. E. (1966) *Astrophys. J.* **143**, 1010.

MENDOZA, E. E. (1968) *Ibid.* **151**, 977.

MORGAN, F. P. (1962) *Q. Report A.A.V.S.O.* No. 25, 45.

NIKOLOV, N. S. and KUNCHEV, P. Z. (1969) *Non-Periodic Phenomena in Variable Stars*, p. 481, Reidel, Dordrecht.

SEGGEWISS, W. and GEYER, E. H. (1969) *Ibid.* 85.

SEGGEWISS, W. (1970) *Astron. J.* **75**, 345.

WEND, R. E. (1960) *Q. Report A.A.V.S.O.* No. 24, 40.

WENZEL, W. (1966) *Mitt. veränderl. Sterne*, **4**, No. 4.

WENZEL, W. (1969) *Non-Periodic Phenomena in Variable Stars*, p. 61, Reidel, Dordrecht.

CHAPTER 3

Spectroscopic characteristics

In general, the spectral classes of the RW Aurigae variables cover the same range as those of the T Orionis stars (types B–K), but while there is this spread of spectral types, we nevertheless find a marked concentration around types G and K as indicated in Table III. As we shall see later, the T Orionis variables tend to cluster into two groups around types B and K.

TABLE III. SPECTRAL TYPES OF RW AURIGAE VARIABLES

	Spectral Type				
	B	A	F	G	K
Number of stars	1	6	15	33	26

THE EMISSION SPECTRUM

In the spectra of these variables, the Balmer lines and the H and K lines of Ca II often appear in emission. The forbidden lines of [S II] are usually present, also in emission, at 4068, 4076, 6717 and 6731 Å. He I is often quite prominent in the spectra of most RW Aurigae variables. In those members with late type spectra, emission lines of [O I] at 6300 and 6363 Å, together with those of [Fe II] and [Ti II], are also observed.

The fact that quite a number of RW Aurigae stars are known which do not show emission lines may be an indication that these members of the group represent T Tauri variables which are older than the rest. Indeed, there is the possibility that emission line activity in the nebular variables as a whole is inversely proportional to their stellar age. A pointer in this direction comes from the positions of the RW Aurigae stars on the Hertzsprung–Russell diagram where they are found either on the main sequence or in the region of the subgiants.

Like the allied T Tauri stars, several of these variables have a fairly well-defined relation between the intensity of the Hα line and the ultraviolet excess (see Kuhi, 1966).

As far as the Balmer emission lines are concerned, the general flatness of the decrement (varying quite smoothly from ~ 0.6 at H8 to ~ 0.2 at H19) and the sharp, rapid rise to a strong Hα, as demonstrated in the profile of this line given in Fig. 15, suggest that here both self-absorption and collisional transitions are important. The effect of the latter has been investigated chiefly by Parker (1964).

As with all of the nebular variables which exhibit emission lines in their spectra, those of the RW Aurigae stars are also found to vary in intensity with changes in the brightness. As we have already seen in the previous chapter, certain small and irregular fluctuations in the light of RW Aurigae, for example, are almost certainly due to variations in the emission line intensity.

The Nebular Variables

Other emission lines have been discovered in several of these stars which have spectra particularly rich in such lines, e.g. RW Aurigae, RU Lupi and V853 Ophiuchi. Here we find emission lines from the permitted transitions of ionized metals and helium, especially those of He I, Na I and Fe II. The lines from forbidden transitions are normally considerably weaker and, in the main, are those due to [O I] and [S II].

THE ABSORPTION SPECTRUM

Very often, there is a late type absorption spectrum underlying the emission one although this is usually obscured by a blue continuum. In addition, the various absorption lines of neutral metals are normally broad and diffuse and tend to be filled in by the underlying continuum. The most likely cause of this line broadening is axial rotation and in this case, these stars are rotating more rapidly than those of similar spectral types on the main sequence.

The broad absorption bands of TiO due to the unresolved M type companion of RW Aurigae are particularly noticeable around minimum light.

ES CARINAE

This star deserves particular mention here. Photographic observations of ES Carinae were made some years ago by Hertzsprung (1925), these showing irregular and isolated variations in brightness with an amplitude of up to 2^m. These variations are often accompanied by more rapid fluctuations of $\sim 0^m.5$ which are typical of the RW Aurigae stars. More recently, this kind of behaviour has been confirmed by Seggewiss and Geyer (1969) from a series of photoelectric determinations.

ES Carinae is situated in a very rich field of the Milky Way at galactic coordinates of $l^{II} = 290°.7, b^{II} = +0°.2$ (1950.0), very close to the galactic plane. It also lies near the centre of the young cluster NGC 3572b, discovered by Schmidt and Santanilla (1964). Distance determinations show, quite conclusively, that this cluster is situated at a greater distance than the nearby cluster NGC 3572 itself and it seems possible that the variable is a member of NGC 3572b. This is certainly indicated by the data given in Fig. 14.

From the known parameters of NGC 3572b, Seggewiss and Geyer have estimated the following absolute visual magnitude and colours for ES Carinae, assuming true membership of this cluster: $M_v = -1^m.65$; $(B-V)_0 = -0^m.14$ and $(U-B)_0 = -0^m.32$. The spectral type therefore corresponds to B7 III.

If this is correct, then ES Carinae has one of the earliest spectral types of all known RW Aurigae variables. As may be seen from Fig. 14, the star is located above the main sequence and it is therefore quite probable that it is still contracting along its evolutionary track towards the main sequence. Another possibility, of course, is that it is now evolving away from the main sequence towards the region of the subgiants although this is somewhat less likely because of its spectral type and the RW Aurigae classification.

R MONOCEROTIS

This variable has been allotted a spectral type of G5$^\pm$ by Jacchia (1945). More recent determinations suggest that it is a nebulous Ae-type star (see Cohen, 1973). The spectrum of this object is dominated by features clearly due to the circumstellar shell and its radiation appears predominantly in the infrared.

The Hα line profile obtained by Dibaj and Esipov (1969) shows an emission shoulder on the major emission peak and this type of profile is intermediate between those of stars such

26

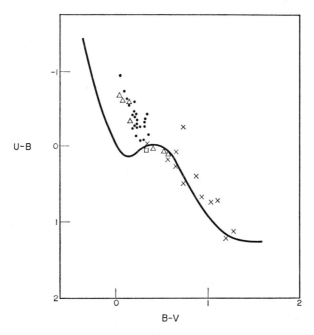

U–B

B–V

Fig. 14. Two-colour diagram of NGC 3572b. The variable is marked as a square, filled circles are cluster members, triangles probable members and crosses are field stars. (After Seggewiss and Geyer.)

as FU Orionis (expanding envelope and self-absorption) and those like V380 Orionis (rotating envelope seen in "pole-on" stars).

SY PHOENICIS

There seems little doubt that this star has an F-type spectrum. The Potsdamer Spektraldurchmusterung gives it as type F4, while the Harvard–Draper Catalogue lists it as F8. The photoelectric observations made in the UBV colour system by Seggewiss and Geyer (1969) also confirm it as an F-type star.

SZ PHOENICIS

The Potsdamer Spektraldurchmusterung gives the spectrum of this star as K4 which again agrees with the photoelectric determinations of Seggewiss and Geyer.

TT PHOENICIS

Although this variable is moderately bright, no spectral type is available in the literature. From its position on the (B–V)/(U–B) diagram, however, it also seems to be a K-type star like the previous variable.

BS VELORUM

This variable is comparatively faint and consequently it is not, perhaps, surprising that no spectral data are available. On the (B–V)/(U–B) diagram of Seggewiss and Geyer it lies very close to the two preceding variables and it would seem to have a similar spectral type.

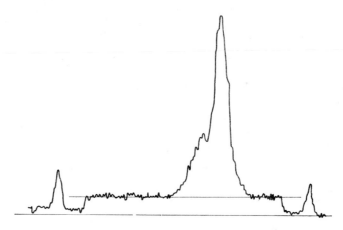

FIG. 15. Hα-line profile for R Monocerotis. (After Dibaj and Esipov.)

Mass Loss Determined Spectroscopically

The variations in the widths of several emission lines in the spectrum of RW Aurigae have been considered by Gahm (1970) in order to determine the mass loss from this star. The results are compared with the model derived by Kuhi (1964) based upon an expanding mass loss envelope around many of the RW Aurigae and T Tauri stars. The mass loss rates found by Gahm indicate that those derived by Kuhi may be appreciably overestimated, particularly for those nebular variables which possess strong emission line spectra. This confirms the conclusion reached by Dibaj and Esipov (1969) that the motions in the atmospheres of these variables are much more complicated than simple expansion.

References

COHEN, M. (1973) *Mon. Not. Roy. astr. Soc.* **161**, 98.
DIBAJ, E. A. and ESIPOV, V. F. (1969) *Non-Periodic Phenomena in Variable Stars*, p. 107, Reidel, Dordrecht.
GAHM, G. F. (1970) *Astron. and Astrophys.* **8**, 73.
HERTZSPRUNG, E. (1925) *Bull. astr. Inst. Netherlands* 210.
JACCHIA, L. (1945) *Astrophys. J.* **102**, 168.
KUHI, L. V. (1964) *Ibid.* **140**, 1465.
KUHI, L. V. (1966) *Publ. astr. Soc. Pacif.* **78**, 430.
PARKER, R. A. R. (1964) *Astrophys. J.* **139**, 208.
SCHMIDT, H. and SANTANILLA, G. D. (1964) *Veröff. Bonn* No. 71.
SEGGEWISS, W. and GEYER, E. H. (1969) *Non-Periodic Phenomena in Variable Stars*, p. 85, Reidel, Dordrecht.

CHAPTER 4

Physical characteristics of RW Aurigae stars

THE RW aurigae variables were first described as a separate class of the nebular variables by Parenago (1933) whose definition restricted them to stars associated with nebulosity and of spectral type G. Many have spectra in which the enhanced lines are weak, indicative of dwarf stars, while emission lines may be either present or absent. As we saw earlier, this latter characteristic appears to have no bearing at all upon their light variations.

A more generalized class was introduced by Hoffmeister (1949). This originally included only stars of spectral type G to M, but was later extended to cover an even wider range of spectral classes.

As shown by Herbig (1962), the RW Aurigae variables form an extremely heterogeneous class which embraces a wide variety of spectral types, masses and luminosities. Consequently, their physical characteristics show a large variation over the group of variables as a whole, ranging from such highly luminous objects as ES Carinae, to dwarf stars with low surface temperatures in the range 4000 to 4500°K (e.g. SZ Phoenicis).

Masses of the RW Aurigae Variables

Any direct determination of the masses of these variables and, indeed, of most of their physical parameters, is complicated by several factors. Being surrounded by dust and gas clouds of various densities and opacities, they seldom show normal stellar spectra and many appear nebulous rather than show stellar disc on long exposure photographs.

In the case of R Monocerotis, for example, Herbig (1968) has shown that most, if not all, of the observed spectral features originate in the circumstellar shell and not in the central star itself. This has a marked effect upon any interpretation of the star. The spectral type of this variable has been given as G5 by Kukarkin and Parenago (1948) whereas it now seems more likely to be a nebulous object of type Be or Ae.

Difficulties may also arise due to the presence of unresolved companions as in the case of RW and UY Aurigae.

This difficulty in classifying the spectra of these stars has been encountered also by Mendoza (1970) who has derived intrinsic colours, bolometric corrections and effective temperatures for certain T Tauri and RW Aurigae variables. From the Hertzsprung–Russell diagram constructed from this data, it would appear that the masses of the RW Aurigae stars fall in the range from $\sim 2\ M_\odot$ to $\sim 3.5\ M_\odot$.

This latter figure is in good agreement with the theoretical calculations made by Larson (1972) for R Monocerotis which is interpreted as an object consisting of a central star with a mass of $\sim 3.5\ M_\odot$ which is surrounded by a protostellar envelope of $\sim 1.0\ M_\odot$. At present, the available evidence, both observational and theoretical, supports the idea that the RW Aurigae stars are, in general, somewhat more massive than the Sun.

The Nebular Variables

The luminosity classes for some of these variables have been determined from which we may obtain an approximate value for their masses (although there will undoubtedly be some deviation from the classical mass-luminosity relation for these gravitationally contracting objects). ES Carinae has a luminosity class III, spectral type B7; RR Tauri has a luminosity class intermediate between II and III, spectral type A2. These two variables, therefore, have approximate masses of 1.5 $M_\odot$ and 3 $M_\odot$ respectively.

Effective Temperatures of the RW Aurigae Variables

From their spectra, and taking into account the various peculiarities which are a common feature of these variables, the effective temperatures of the RW Aurigae stars vary from $\sim 1.5 \times 10^{4}°$K (ES Carinae, V852 Ophiuchi, RR Tauri) to as low as $4.0 \times 10^{3}°$K (V853 Ophiuchi, V426 Orionis, SZ Phoenicis).

The most comprehensively studied object of this class is R Monocerotis. According to Herbig (1968) the nebular spectrum obscures much of the stellar radiation but there are fairly strong absorption lines which seem to originate in the stellar atmosphere and indicate an effective temperature of $\sim 10^{4}°$K. This corresponds to a spectral class of about A2; some confirmation of this being the weakness of the helium lines.

THE SURROUNDING CIRCUMSTELLAR SHELL

The infrared observations made of several of these variables give an estimate of the grain temperature in the circumstellar dust shells which enshroud the central core. These are, of course, considerably lower than the effective temperatures of the stars. For most of these variables, the infrared continuum peaks at $\sim 3 \mu$ indicate a grain temperature of $\sim 600°$K.

For R Monocerotis, the grain temperature measured in this way is slightly higher at $\sim 850°$K. From the close similarity of the grain temperatures for a fairly wide range of luminosities of the stars, it appears that this may be determined by the nature of the dust grains rather than by the properties of the central objects. Several recent estimates of the temperatures in the surrounding dust clouds have been made by Low (1970).

Luminosities of the RW Aurigae Variables

Obscuration due to the circumstellar dust shells around these objects makes it difficult to determine precise luminosities of the stellar cores merely from observation. Photometric and spectroscopic determinations by Low (1970) and Low et al. (1970) suggest bolometric luminosities for these variables of between 70 $L_\odot$ and 480 $L_\odot$.

We may compare these values with the theoretical properties of the stellar cores when they approach, or reach, the main sequence assuming the non-homologous collapse of the protostellar cloud. The above values agree very well with those predicted for protostars having masses between 2.5 $M_\odot$ and 5.0 $M_\odot$. On this theory, for example, R Monocerotis has a predicted luminosity of $\sim 870 L_\odot$.

Stellar Radii

The radii of the majority of the RW Aurigae stars lie between 2 $R_\odot$ and 3 $R_\odot$. From their spectral types and/or luminosity class, we might expect stars such as V852 Ophiuchi and RR Tauri to be larger than this although no estimate of their radii is available.

The theoretical radii of the protostars before they become visible in the optical region depends largely upon whether the protostellar cloud collapses homologously as suggested by Hayashi (1961), Hayashi *et al.* (1962) and Hayashi and Nakano (1965) or non-homologously as examined by Larson (1969, 1972a).

In the former case such protostars will come onto their "Hayashi" tracks with very large radii of the order of 100–$250\ R_\odot$. The radius will then decrease as the protostar evolves down its track towards the main sequence. In the case of the RW Aurigae variables, of course, there will be no "Hayashi" track since their masses are much too large for them to be fully convective and they will first appear close to the main sequence.

According to the theory of non-homologous collapse of the cloud, the radius of the proto-star varies little with mass by the time it is contracting towards the main sequence in radiative equilibrium. Such radii vary from $1.8\ R_\odot$ for a mass of $0.25\ M_\odot$, to $3.7\ R_\odot$ for a mass of $10\ M_\odot$.

Stellar Motions

Perhaps the most significant feature of the stellar motions of these variables, in common with those of the T Orionis and T Tauri stars, is their small value. Those which have been determined so far, lie between 15 and 30 km/sec. This clearly shows that these variables are not field stars which are merely moving through the nebulae with which they are associated, providing further confirmatory evidence that they are physically connected with these nebulae.

During the later stages of the evolution of a rotating protostellar cloud, Larson (1972b) has shown that, so long as magnetic torques are insufficiently strong enough to damp out rotational motions on a time scale less than the free-fall time, the deviations from spherical symmetry will become magnified during the ensuing collapse. The most likely result will be the eventual formation of two or more stellar cores orbiting each other.

While fragmentation appears to be inhibited in a non-rotating cloud, rotation is probably the primary promotor of fragmentation. As a result, we would expect to find a number of binary, or multiple, systems in which one, or more, components are RW Aurigae variables. This is borne out by observation.

Thus we have RW Aurigae itself which has an M type companion (which may itself be variable in brightness to a certain degree) and stars such as UY Aurigae, the companion of which also possesses an emission spectrum although it appears to be constant in brightness. As we shall see, we find similar systems among the T Orionis and T Tauri variables.

References

HAYASHI, C. (1961) *Publ. astr. Soc. Japan* **15**, 450.
HAYASHI, C., HOSHI, R. and SUGIMOTO, D. (1962) *Prog. Theor. Phys. Suppl.* No. 22.
HAYASHI, C. and NAKANO, T. (1965) *Prog. Theor. Phys.* **34**, 754.
HERBIG, G. H. (1962) *Adv. Astr. Astrophys.* **1**, 47.
HERBIG, G. H. (1968) *Astrophys. J.* **152**, 439.
HOFFMEISTER, C. (1949) *A.N.* **278**, 24.
KUKARKIN, B. V. and PARENAGO, P. P. (1948) *General Catalogue of Variable Stars*, First Ed., Moscow.
LARSON, R. B. (1969) *Mon. Not. Roy. astr. Soc.* **145**, 271.
LARSON, R. B. (1972a) *Ibid.* **157**, 121.

The Nebular Variables

LARSON, R. B. (1972b) *Ibid.* **156**, 437.
LOW, F. J. (1970) *Mem. Soc. Roy. Sci. Liége* **19**, 281.
LOW, F. J., JOHNSON, H. L., KLEINMANN, D. E., LATHAM, A. S. and GEISEL, S. L. (1970) *Astrophys. J.* **160**, 531.
MENDOZA, E. E. (1970) *Mem. Soc. Roy. Sci. Liége* **19**, 319.
PARENAGO, P. P. (1933) *Peremennye Zvezdy* **4**, 222.

CHAPTER 5

Spatial distribution

FROM the numerous photographic and spectroscopic surveys that have been carried out we are able to draw certain conclusions regarding both the overall, and the particular, distribution of the nebular variables. The majority of those known at the present time lie within a radius of 100 pc of the Sun and of these, almost all are situated in a fairly narrow band on either side of the galactic equator. Scarcely any have been discovered in galactic latitudes greater than $\pm 20°$. Within this region we find large numbers of diffuse nebulae, both bright and dark, and it is in the vicinity of these that the nebular variables tend to congregate. In the case of RW Aurigae, however, as Wenzel (1969) has pointed out, this star lies at a distance of some 2° from the nearest dark cloud.

This latter relation between the nebular variables and these gas and dust clouds is, as we shall see, a necessary one since all of these stars are contracting objects, still in their pre-main sequence stages of evolution.

In spite of the obviously close relationship among the RW Aurigae, T Orionis and T Tauri variables, we do find some differences in their spatial distributions within the band just mentioned. In Chapters 11 and 17, it will be shown that the T Orionis stars demonstrate a distinct tendency to concentrate within the Orion complex and the T Tauri variables congregate both here, in Monoceros (near NGC 2264) and among the Taurus dark clouds. The RW Aurigae stars, however, appear to be more widely, and evenly, scattered along the region of the galactic equator.

Table IV illustrates the approximate galactic coordinates of several of the known associations of RW Aurigae variables.

TABLE IV. GALACTIC COORDINATES OF RW AURIGAE ASSOCIATIONS

Constellation	Galactic coordinates*	
	l^{II}	b^{II}
Serpens	3°	+ 8°
Ophiuchus	7°	+ 9°
Aquila	9°	+ 3°
Delphinus	28°	−15°
Perseus	103°	− 8°
Auriga	137°	− 1°
Taurus	148°	− 2°
Monoceros	179°	+ 3°
Monoceros	187°	− 2°
Chamaeleon	267°	−17°
Lupus	307°	+13°
Ophiuchus	358°	+13°

*Epoch 1950.0.

The Nebular Variables

RW Aurigae Variables in Specific Regions

SERPENS REGION

This region contains a small number of RW Aurigae variables. The two brightest representatives are BK Serpentis (14.3–16.7 m_{pg}) and BQ Serpentis (11.0–11.9 m_{pg}) both of which have been described by Ahnert (1943).

OPHIUCHUS REGION

Several regions of nebulosity are located in this constellation which contain numerous faint nebular variables. At the present time, few comprehensive light curves for these stars are available for study. Two areas containing RW Aurigae variables, together with T Tauri stars, have been examined in detail within the Milky Way to the north of the Aquila Rift.

Typical representatives of the RW Aurigae variables for which some details of their light curves are available, are V426 Ophiuchi (12.2–13.8 m_{pg}) described by Hoffmeister (1938) and V637 Ophiuchi (13.6–16.0 m_{pg}) discovered during an extensive photographic survey of this region by Ahnert (1943).

AQUILA REGION

Several nebular variables of all three types have been found in the dark nebulosities in this area, particularly in the region around NGC 6709 and NGC 6724. Most of these variables were discovered by Ahnert (1943), including such stars as V480 Aquilae (44.0–15.9 m_{pg}) and V489 Aquilae (13.1–14.8 m_{pg}).

A more recent survey by Götz (1955b) has resulted in the discovery of further members of this group, e.g. V800 Aquilae (12.8–15.7 m_{pg}). Very few details of the light curves and spectra of these faint variables are available.

DELPHINUS REGION

A small number of nebular variables in this region have been described by Hoffmeister (1936). Although few observations have been obtained, from their light variations they appear to be either RW Aurigae or T Tauri variables. BZ Delphini (13.6–14.3 m_{pg}) is typical of these stars. Some caution should be exercised over the apparently small amplitude of this, and other faint variables, since insufficient data are available at the moment to enable accurate estimates of the extreme range to be determined.

PERSEUS REGION

The nebular variables located in the dark clouds found in this constellation are generally very faint objects such as DN Persei (14.8–16.2 m_{pg}) discovered by Hoffmeister (1944). As a result, their light variations have been only fragmentarily investigated and some difficulty is encountered in assigning them unambiguously to any of the three classes. Much more observational data are required for these stars including spectroscopic determinations.

AURIGA–TAURUS REGION

Large numbers of RW Aurigae and T Tauri variables are found in the dark clouds of Auriga which extend into the neighbouring constellation of Taurus. The prototype star is located in this region, together with the bright variables UY Aurigae and TRR Tauri.

Among the fainter nebular variables that have been discovered, DY Aurigae (13.3–

34

16.7 m$_{pg}$), which has been described by Ahnert (1941), appears to be a genuine RW Aurigae star.

MONOCEROS REGION

A large concentration of nebular variables is found here, particularly in the region of NGC 2264. Most of these were discovered by Herbig (1954) during an objective prism survey of this region for stars showing Hα emission spectra. Subsequent photographic work by Wenzel (1955a, 1955b) proved many of these stars to be irregularly variable, often with amplitudes up to 2^m.

Light curves have been obtained for several of these variables and these indicate that, while the majority are undoubtedly T Tauri stars, a small number appear to belong to the RW Aurigae class. One of the variables found in this region, PZ Monocerotis (spectral type dK2), is noteworthy in that, unlike the majority of these stars it shows no absorption line due to Li I at 6708 Å. This is a distinction it shares with the T Tauri variable CQ Tauri (spectral type dF2).

CHAMAELEON REGION

T Chamaeleontis is one of the brighter and more comprehensively observed of the RW Aurigae variables having been closely followed for many decades, particularly by the Variable Star Section of the New Zealand Astronomical Association (Bateson, 1950 et seq.). The star has also been fully described by Hoffmeister (1943). Several other members of this group are located in the T-association close to this variable.

Recently, Mendoza (1972) has reported results of a survey for Hα emission stars in a nearby region to the north of T Chamaeleontis which covers an area of 1°.3 × 1°.6. Twenty-four objects which show Hα in emission have been discovered with photographic magnitudes between 12^m and 17^m. Slit spectra and Schmidt (ultraviolet and red) spectra indicate that all of these objects belong to this nearby T-association. Since the majority of these new variables are extremely faint objects, even at maximum brightness, very few details of their light curves are, as yet, available. It seems probable, on the present data, that the majority belong to the T Tauri class, although a few may be RW Aurigae stars.

LUPUS REGION

Two bright nebular variables have been discovered and described by Hoffmeister (1943) which, on the basis of their typical light variations, are RW Aurigae stars. These are RU Lupi (9.3–13.2 m$_{pg}$) and RY Lupi (9.9–13.0 m$_{pg}$). In contrast to other variables of the RW Aurigae class, the latter star exhibits scarcely any fluctuations at all around maximum light according to Kukarkin and Parenago (1948).

FL 33 CYGNI FIELD

In addition to the above fairly well-defined regions, isolated RW Aurigae variables have been discovered during the numerous photographic surveys of selected fields in the Milky Way. Götz (1955a) has reported the discovery of a faint nebular variable of the RW Aurigae type in this particular field on plates taken with the 400/1600 mm astrograph of the Sonneberg Observatory. The amplitude of this star is quite small (range 12.9–15.3 m$_{pg}$).

BETA CASSIOPEIAE FIELD

Among the large number of variables in this field which have been examined by Wenzel

The Nebular Variables

(1955c) are a handful of RW Aurigae stars with small amplitudes $\sim 0^m.7$. Further observations of these variables are required, however, before they can be definitely assigned to this class.

PHI CASSIOPEIAE FIELD

During a photographic survey of the nebulous region around ϕ Cassiopeiae, Götz (1955a) discovered two further RW Aurigae variables. S 3867 Cassiopeiae has quite a large amplitude (13.7–17.6 m_{pg}) while S 3884 Persei has a much smaller range (11.7–12.8 m_{pg}). More observational data are necessary before we can rule out the possibility of the latter being a T Tauri variable.

References

AHNERT, P. (1941) *Kleinere Veröff. der Univ. Berlin-Babelsberg*, No. 24.
AHNERT, P. (1943) *Ibid*. No. 28.
BATESON, F. M. (1950 *et seq*.) *Q. Rep. Variable Star Section*, New Zealand Astronomical Association.
GÖTZ, W. (1955a) *Mitt. veränderl. Sterne* No. 181.
GÖTZ, W. (1955b) *Ibid*. No. 183.
HERBIG, G. H. (1954) *Astrophys. J.* **119**, 483.
HOFFMEISTER, C. (1936) *Thesis*, Univ. Berlin.
HOFFMEISTER, C. (1938) *Kleinere Veröff. der Univ. Berlin-Babelsberg*, No. 19.
HOFFMEISTER, C. (1943) *Ibid*. No. 27.
HOFFMEISTER, C. (1944) *Mitt. veränderl. Sterne* No. 64.
KUKARKIN, B. V. and PARENAGO, P. P. (1948) *General Catalogue of Variable Stars*, p. 400, First Ed., Moscow.
MENDOZA, E. E. (1972) *Publ. astr. Soc. Pacif.* **84**, 641.
WENZEL, W. (1955a) *Mitt. veränderl. Sterne* No. 190.
WENZEL, W. (1955b) *Ibid*. No. 193.
WENZEL, W. (1955c) *Ibid*. No. 174.
WENZEL, W. (1969) *Non-Periodic Phenomena in Variable Stars*, p. 63, Reidel, Dordrecht.

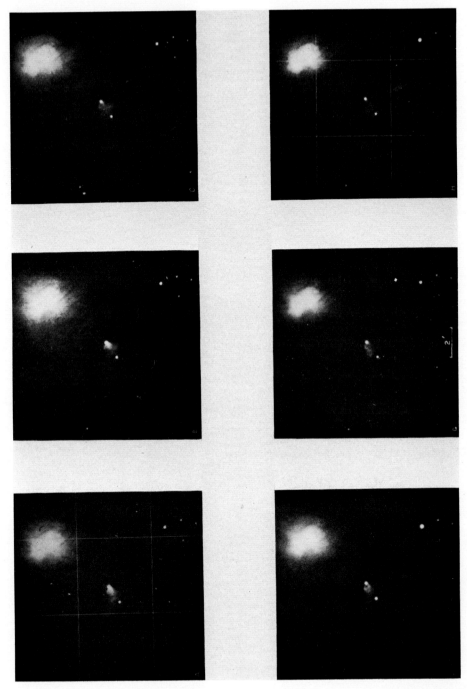

PLATE II. Variations in NGC 2261 associated with R Monocerotis. (Courtesy Lowell Observatory.)

CHAPTER 6

Interaction with nebulosity

VIRTUALLY all of the contracting, pre-main sequence variables, of which the RW Aurigae stars form a group, are in physical contact with interstellar matter. The light variations are determined, in part at least, by the local behaviour of the circumstellar dust shells in which they are embedded while, on the other hand, these shells possess a definite connection with the larger interstellar clouds.

We know that stellar formation in certain regions can continue for a relatively long period (in the Taurus and Orion associations, for example) and consequently we should expect to find variables of different stellar ages within the same association. The real age of a variable plays an important role in the various spheres of activity going on between star, circumstellar shell and interstellar matter since, on average, this physical connection decreases with increasing stellar age.

The nebulae associated with several RW Aurigae variables, e.g. R Monocerotis and RR Tauri, belong to the small group of peculiarly-shaped nebulosities commonly classed as cometary nebulae. R Monocerotis is connected with the conical nebula NGC 2261, while the morphology of the nebula near RR Tauri places it in the comma-like group as defined by Dibaj (1963). The comparable ages of both star and associated nebula have been demonstrated by several workers including Ambartsumian (1957), Poveda (1965) and Dibaj (1969). All of these determinations point to a definite relation between the cometary nebulae and early stellar formation.

The spectra of the cometary nebula, in particular NGC 2261 around R Monocerotis, have been investigated by Kazarian and Khachikian (1972). All of these nebular spectra show a large number (~ 100) of emission and absorption lines which have been identified with H I, Fe I, Fe II, Ti I, Ti II, Cr II, Ca II and Sc II. The forbidden line of [O II] at 3727 Å has been positively identified in NGC 2261.

In this context, it is significant that the emission line of [O II] is first observed $\sim 40''$ to the north of the stellar nucleus with the intensity increasing gradually towards the periphery of the nebula. The emission lines of Hα and Hβ are also present to the south of R Monocerotis. From this observation, Kazarian and Khachikian have concluded that there must be some energy source within the nebula itself which generates light quanta. So far, the nature of this source has not been identified. The continuum of the nebular spectrum appears to be of a reflection type.

The infrared observations made by Cohen (1973) of several RW Aurigae stars, including RW and UY Aurigae, R Monocerotis, CO Orionis and RR Tauri show that all possess large infrared fluxes as far as 11μ. R Monocerotis, in particular, is very bright at wavelengths up to 20μ. In most of these stars, if not all of them, therefore, the major proportion of their luminosity is found in the infrared beyond 2μ. From this we may assume that they are still

The Nebular Variables

shrouded by extensive dust clouds which absorb much of the optical radiation from the high luminosity stars embedded within them.

Some of this excess infrared emission may be due to free–free emission which has been found in a number of these young stars by Breger and Dyck (1972). Such emission can extend as far as 11μ and could arise from the very hot gas (temperatures up to $10^{7}°K$) just inside the shock front surrounding the central core. Some confirmation of the presence of continuous free–free emission comes from the fact that a number of stars with infrared excesses also possess appreciable polarization at wavelengths of $\sim 0.75\mu$ which may be due to electron scattering.

The greater part of the infrared radiation, however, must be due to thermal emission from the grains within the circumstellar dust shells and certainly this will be responsible for the long wavelength brightness of these objects.

From all of these observations, we may conclude that in the majority of RW Aurigae variables, there is still sufficient dust present for the infall of material onto the evolving star to be one of the dominant processes occurring in the vicinity of these objects.

Effects of Infalling Material

From theoretical considerations, it appears that stars in the mass range of 3.5 $M_\odot$ to 5.0 $M_\odot$, e.g. R Monocerotis, have settled into radiative equilibrium by the time they become visible and approach the main sequence. Consequently, these objects are now contracting and evolving along a radiative track although the star is still accreting mass from the surrounding cloud. Since these two effects are in opposition to each other, the radius remains virtually constant even though the mass is increasing.

In some of the fainter T Orionis and T Tauri variables, Walker (1961, 1963, 1964, 1969) has observed absorption components on the emission lines that are red-shifted by 150–300 km/sec and interpreted these as evidence of infalling material in these stars. So far, such Doppler-shifted absorption lines have not been reported for the RW Aurigae variables where, if an excess of such material is still present, we would expect to find them. A possible explanation of this anomaly is that the spectra of the central stars in most of the RW Aurigae systems are so dominated by the spectra of the surrounding clouds that such peculiarities are both difficult to observe and measure.

However, the infalling material in these higher-mass stars has a relatively high velocity and kinetic energy. This will be sufficient to offset the radiative cooling of the core to a certain extent maintaining, or even increasing, the temperature just inside the surrounding shock front. This can account for the gradual increase in the luminosity and bolometric temperature as these variables approach the main sequence (Fig. 18). We would therefore expect these stars to be more luminous than corresponding main sequence stars of the same spectral type. In Chapter 4 we saw that this is, indeed, the case.

The Shock Front (Larson, 1969)

In the treatment of the shock front, a set of equations analogous to the normal adiabatic shock jump equations is required so that we can relate the values of the flow variables at two points, one just inside the shock front and the other just outside.

The conservation of mass relation for the shock region is

$$\rho_1 u_1 = \rho_2 u_2 \tag{1}$$

where the subscript 1 denotes quantities at a chosen point inside the shock front and subscript 2, similar quantities at a point outside. It is assumed that there is a steady flow through the shock front and the velocity of the shock front is extremely small in comparison with the velocity of the infalling material.

From the energy inflow and outflow rates across the two surfaces of the shock front, it is then possible to derive an energy conservation relation for this shock region. The ordinary mechanical transport rates consist of (a) an energy inflow of $\rho_2 |u_2|(H_2+\frac{1}{2}u_2{}^2)$ and (b) an energy outflow of $\rho_1 |u_1|(H_1+\frac{1}{2}u_1{}^2)$ where H is the specific enthalpy (Liepmann and Roshko, 1957).

The radiative energy transport is also important here. The energy input into the shock front is the outward energy F_1 from the interior of the stellar core and this may be calculated since it depends in a known manner upon the temperature gradient in the surface layers of the core. Similarly, the outward energy flux F_2 emitted from the shock front depends upon the temperature distribution inside the shock front. In the present context it is desirable to relate F_2 to a single parameter T_1 which is the limiting temperature inside the shock front by the time radiative cooling has become negligible.

Now the temperature immediately inside the shock jump, at this stage, is extremely high ($\sim 10^{7\circ}$K) due almost entirely to the high kinetic energy associated with the infalling material. Consequently, it is far from easy to relate F_2 to T_1. However, since the effective temperature T_e is defined by

$$F_2 = \sigma T_e{}^4 \tag{2}$$

and it can be shown that, no matter how high the peak temperature inside the shock jump may be,

$$T_1 < T_e < 1.28 T_1 \tag{3}$$

we may adopt, as a suitable approximation

$$T_e = T_1. \tag{4}$$

Collecting together the various energy gain and loss terms, we have the necessary energy conservation relation for the shock region, viz.

$$\rho_1 |u_1|(H_1+\tfrac{1}{2}u_1{}^2)+\sigma T_1{}^4 = \rho_2 |u_2|(H_2+\tfrac{1}{2}u_2{}^2)+F_1. \tag{5}$$

We must now consider the shock momentum equation for the determination of the pressure P_1 inside the shock front. By the time the central core has evolved to the point where it is, to all intents and purposes, a star, the region inside the shock front represents a stellar atmosphere. We therefore choose P_1 at a point where the optical depth inside the shock front is of the order of unity. On the basis of the Eddington approximation, the photospheric boundary (where the actual surface temperature is equal to the effective temperature) occurs at an optical depth of 2/3.

Integrating the hydrostatic equilibrium equation

$$\frac{\mathrm{d}P}{\mathrm{d}\tau} = \frac{g}{\kappa} \tag{6}$$

between $\tau = 0$ and $\tau = 2/3$, and making the assumption that $\kappa \propto P^{\frac{1}{2}}$ it can be shown that the photospheric pressure P_1 exceeds the pressure immediately inside the shock jump by an amount approximately equal to g/κ_1. This term must therefore be added to the normal

The Nebular Variables

adiabatic shock momentum equation to yield the required approximate relation

$$P_1 + \rho_1 u_1{}^2 = P_2 + \rho_2 u_2{}^2 + \frac{g}{\kappa_1}. \tag{7}$$

All of the variables on the two sides of the shock front may therefore be related by means of equations (1), (5) and (7).

References

AMBARTSUMIAN, V. A. (1957) *Non-Stable Stars*, Cambridge Univ. Press.
BREGER, M. and DYCK, H. M. (1972) *Astrophys. J.* **175**, 127.
COHEN, M. (1973) *Mon. Not. Roy. astr. Soc.*, **161**, 97.
DIBAJ, E. A. (1963) *Sov. Astr. A. J.* **7**, 606.
DIBAJ, E. A. (1969) *Astrophysics* **5**, 115, Faraday Press, New York.
KAZARIAN, M. A. and KHACHIKIAN, E. YE. (1972) *Astrofizika* **8**, 17.
LARSON, R. B. (1969) *Mon. Not. Roy. astr. Soc.* **145**, 289.
LIEPMANN, H. W. and ROSHKO, A. (1957) *Elements of Gas Dynamics*, Ch. 2, Wiley, New York.
POVEDA, A. (1965) *Bol. Obs. Tonantzintla Tacubaya* **4**, 15.
WALKER, M. F. (1961) *Comptes Rendus* **253**, 383.
WALKER, M. F. (1963) *Astrophys. J.* **68**, 298.
WALKER, M. F. (1964) *Roy. Obs. Bull.* No. 82, 69.
WALKER, M. F. (1969) *Non-Periodic Phenomena in Variable Stars*, p. 103, Reidel, Dordrecht.

CHAPTER 7

Evolutionary characteristics

By now it will have become apparent that nearly all of the contracting pre-main sequence stars are directly connected with interstellar matter in the form of dark or bright nebulosity. Because of this close association, and their position above the main sequence on the Hertzsprung–Russell diagram, it has long been believed that the RW Aurigae variables (in common with the allied T Orionis and T Tauri stars) are very young objects surrounded, to varying extents, by the remains of the protostellar envelope from which they have condensed.

There is some, admittedly tenuous, evidence that our present division of these variables into the above three classes may be, partially at least, attributed to their masses. From theoretical calculations made by Larson (1972), R Monocerotis (RW Aurigae type) can be interpreted as a protostar with a total mass of ~ 4–$5\ M_\odot$; R Coronae Australis (T Orionis type) has a total mass of $\sim 2.5\ M_\odot$, and the typical T Tauri variables have total masses in the range of 0.8–1.5 $M_\odot$.

As expected, we find some scatter of positions of these variables on the Hertzsprung–Russell diagram above the main sequence and almost certainly this arises from various effects such as variations in the amount of circumstellar absorption and differences in the initial conditions prior to stellar formation. Evidence for the presence of circumstellar shells around the pre-main sequence stars which accounts for some of this scatter has been found by several observers including Breger (1972) and Strom *et al.* (1971, 1972a, 1972b).

We must also take into account the age spread of these variables which are located in young clusters such as those in the Orion Nebula and the Auriga-Taurus dark clouds. Larson (1972) has shown that the time required for the formation of a stellar nucleus from the protostellar cloud is of the order of a few times the free fall time (the time of collapse of the cloud under gravity). This time is of a similar order to the variation we find in formation time brought about by differences in the initial and boundary conditions for the cloud.

Now the free fall time is dependent upon the density of the protostellar cloud. Assuming the densities in young clusters (gas, dust and stars), and in the interstellar cloud from which the cluster is formed, to be approximately the same, we can estimate the free fall time directly from observation.

Menon (1966) has estimated that the density of the Orion cluster varies from $\sim 10^{-19}$ g cm^{-3} at the centre to $\sim 10^{-22}$ g cm^{-3} in the outermost regions. This yields a free fall time ranging between 2×10^5 and 6×10^6 years. We therefore have a spread of ages of at least 10^7 years. Indeed, it is quite probable that this figure may be even higher if any accretion of material from the region surrounding the cloud takes place.

The BM Andromedae Group

Some additional confirmation of this age spread comes from the observational work

The Nebular Variables

of Aveni and Hunter (1970) who have studied the BM Andromedae complex. The T Tauri variable BM Andromedae is immersed in the southern tip of an isolated tongue of nebulosity which is situated ~ 80 pc above the galactic plane. Ten probable members of this group, together with several possible outlying members, have been discovered from their photometric and spectroscopic characteristics.

Since stars later than spectral type A0 have not yet contracted onto the main sequence in this region, the age of the complex is $\sim 10^7$ years. The total mass of the group (including interstellar material) appears to be approximately 60 $M_\odot$. In view of its isolation from other early-type clusters and associations, Aveni and Hunter have suggested that the BM Andromedae complex is an example of one of the smaller-mass primary condensations where star formation is currently going on.

It therefore seems likely that, whereas the other effects mentioned earlier contribute to the observed scatter on the Hertzsprung–Russell diagram, the age spread is the major contributor.

The Protostellar Cloud

In the formulation of the various models for pre-main sequence stars and their evolution, certain assumptions have to be made regarding the initial and boundary conditions of the protostellar cloud. Clearly, the initial temperature is one of the more important parameters since this governs, through the Jeans criterion, the density at which gravitational collapse begins for the condensation of a given mass.

In the past, estimates as high as 100°K have been made of the temperatures found in the dense dark clouds where star formation appears to occur. More recent work, particularly observations made by Heiles (1971), now suggests that much lower temperatures of between 5 and 30°K prevail in these clouds. Theoretical considerations, too, predict temperatures in the range 5 to 20°K according to Field (1970) and Hayashi (1966). This is especially so for clouds with mean densities in the range we are interested in here, namely in excess of $\sim 10^2$ atoms cm^{-3}.

Both heating and cooling effects will be present during the collapse of such a cloud. Heating will come from the compression of the material during gravitational collapse at approximately free fall rate. Cooling will be brought about by the transfer of energy by collisions of gas molecules with grains of dust followed by radiative cooling of the dust grains. One may therefore assume that both initially and during the optically thin stages of collapse, the protostellar cloud will be isothermal at a temperature of $\sim 10°$K.

Assumption must also be made concerning the opacity and composition of the material within the cloud. Observation provides evidence for the presence of dust grains (graphite or silicates) and Gaustad (1963) has shown that, so long as dust grains are present, they are by far the most dominant source of opacity. Larson (1969, 1972) in his study of the non-homologous collapse of such a cloud, has assumed a constant value for the Rosseland mean opacity of 0.15 cm^2 g^{-1} and also that all of the hydrogen is present in the molecular form.

In the simplest case, the collapsing cloud may be assumed to start from rest and possess a uniform density distribution, the corresponding boundary condition being that the outer boundary of the cloud remains fixed in space with a constant radius.

The maximum radius for gravitational collapse has been found by Larson (1969) to be given by

$$R_{\text{max}} = 0.46 \frac{GM}{\mathscr{R}T} \tag{1}$$

where G is the gravitational constant, M is the mass of the cloud and $\mathscr{R}$ is the gas constant. There are, however, certain difficulties found if this radius is adopted. For radii only slightly greater than R_{max} it may be shown that although collapse occurs, the cloud then rebounds and does not collapse into a stellar core. Furthermore, any small errors in the calculations can also lead to the "rebound" of the model and prevent stellar formation. Larson (1972) therefore adopts a slightly smaller radius given by

$$R = 0.41 \, \frac{GM}{\mathscr{R}T}. \tag{2}$$

McCrea (1957) has already shown that such a radius is small enough to ensure that gravitational collapse occurs from arbitrary starting conditions. Several investigators, notably Penston (1966) and Bodenheimer and Sweigart (1968) have shown that the major features of the collapse are altered but little by changes in the initial and boundary conditions of the protostellar cloud.

With regard to the initial mass of the protostellar cloud, Wright (1970) has traced possible evolutionary paths for protostars from diffuse clouds of interstellar gas with a mean density of 10^{-23} g cm^{-3}, through fragmentation, to the time when nuclear burning commences and an average density of ~ 1 g cm^{-3} is attained. From these results it is concluded that a minimum mass of 10^{32} g is required for such a cloud to contract to a pre-main sequence star.

At this point we must consider two different theories concerning the collapse of a protostellar cloud.

Homologous Collapse

Models based upon the theory of the homologous collapse of a gas and dust cloud have been worked out primarily by Hayashi *et al.* (1962), Hayashi and Nakano (1965) and Hayashi (1966), while models for rapidly rotating pre-main sequence stars assuming homologous collapse have been put forward by Moss (1973). Since the effect of rotation upon the evolution of a pre-main sequence star has been investigated primarily for objects having masses of $1 \, M_\odot$ or less, this subject will be treated later, in detail, in Chapter 19 when we come to discuss the T Tauri variables. Here we shall concern ourselves with non-rotating models.

On the assumption that a large mass of gas collapses homologously (or consists of polytropic density distributions), the quasi-static configuration when it may first be described as a star will, of necessity, possess both a large radius and high luminosity. If the mass of this object is not too large, i.e. of the order of $1 \, M_\odot$, then it will be fully convective as shown by Hayashi (1961). On the Hertzsprung–Russell diagram it will then follow a track such as that shown in Fig. 16, descending almost vertically to the right of the main sequence.

The combination of high luminosity ($\sim 300 \, L_\odot$ for a star of $1 \, M_\odot$) and high opacity during the contraction phase down the "Hayashi" track demands that the star be fully convective since otherwise it would be impossible for the energy to be transported sufficiently rapidly to the surface. In other words, such stars cannot be in hydrostatic equilibrium (surface temperature below 3000°K) in this region. As we shall see, stars with masses in excess of $\sim 2 \, M_\odot$ cannot remain fully convective during their pre-main sequence lifetimes and consequently have no "Hayashi" phase.

Non-Homologous Collapse

Larson (1972) has calculated evolutionary models of spherical protostars for a range of

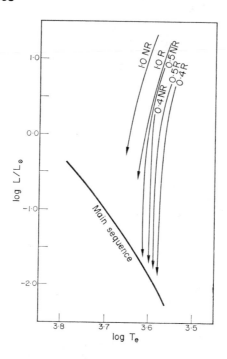

FIG. 16. Hertzsprung–Russell diagram for rotating and non-rotating models assuming homologous collapse. The constant mass tracks are labelled with the mass in solar units and R is for rotating and NR for non-rotating models. (After Moss.)

masses from 0.25 $M_\odot$ to 10 $M_\odot$ on the basis of the non-homologous collapse of the proto-stellar cloud. On the rather tacit assumption that the RW Aurigae variables have the largest masses of the three classes into which we have divided the main nebular variables, we shall here be considering only those models with a total mass of 5 $M_\odot$. Models for protostars with smaller masses will be discussed in Chapters 13 and 19.

EVOLUTION OF THE STELLAR CORE

Larson (1969) has shown that there are seven phases which may be distinguished during the formation of a star from the protostellar cloud and that these are generally very similar for the entire range of masses from 0.25 $M_\odot$ to 10 $M_\odot$. Following his treatment of the problem we shall discuss each of these in turn and finally compare the end result with observation.

THE INITIAL ISOTHERMAL PHASE

Since the assumption has been made that there are no pressure gradients within the protostellar cloud, the entire cloud will begin its collapse in free fall. Once this has begun, however, a pressure gradient will form in the outermost regions since the density will rise in the centre and fall near the boundary. One effect of this pressure gradient will be that there will be a retardation from free fall in this region. Consequently, the density in the centre will increase more rapidly than in the outer layers of the cloud.

This non-homologous character of the collapse appears to take place regardless of the initial and boundary conditions. We thus have the quite rapid formation of a density peak at the centre of the cloud while the remainder will change very little during this phase.

FORMATION OF AN OPAQUE CORE

By now the density at the centre will have risen from the initial value of $\sim 10^{-19}$ to $\sim 10^{-13}$ g cm^{-3}. Once this happens, this small region becomes opaque. It is no longer possible for the heat generated by the collapse in this region to radiate freely away and we have an approximately adiabatic compression. This results in a further rapid rise in both temperature and pressure which soon causes the collapse in the centre to cease.

This small central core is now in almost hydrostatic equilibrium, but outside, the infalling material (still in virtual isothermal conditions) sets up a shock front as it meets the core boundary. The temperature and density of the core are now $\sim 170°$K and 2×10^{10} g cm^{-3} respectively.

The continual infall of material brings about an increase in the mass of the core, but because of loss of energy by radiation from the surface of the core, the radius continues to decrease.

FORMATION OF THE SECOND CORE

Eventually, when the mass of the core has approximately doubled, with a resultant halving of its radius, the central temperature reaches the point where it is sufficient to dissociate the hydrogen molecules (about 2000°K). The material at the centre of the core now collapses dynamically because of an instability which is set up due to the ratio of the specific heats falling below the critical value of 4/3. On a very short time scale, the central density again increases sharply.

We therefore have two concentric cores with density and velocity distributions similar to those illustrated in Fig. 17.

EXPANSION OF THE STELLAR CORE

Shortly after the formation of the second core, depletion of the material in the first core takes place with a resultant sharp drop in the density of the infalling material. While the temperature decreases only slightly, there is an accompanying rather drastic fall in the density just inside the shock front. The core therefore expands with the radius increasing roughly proportionally to its mass. For a star such as R Monocerotis, the maximum radius at this stage of its evolution would be $\sim 50\ R_\odot$, the mass of the core being $\sim 5 \times 10^{-2}\ M_\odot$.

Here we must consider the effect of increasing ionization which now occurs with the continued rise in the central temperature and density. Calculation shows that this does not bring about any further collapse of the core since the ionization is still not sufficiently high to reduce γ, the ratio of the specific heats, below 4/3.

COOLING AND CONTRACTION OF THE CORE

This phase begins once the density outside the shock front falls below a critical value of $\sim 10^{-8}$ g cm^{-3}. At this density, the opacity of the infalling material is reduced sufficiently for the radiative energy from the shock front to pass through the cloud and out into space. Because of this radiative loss of energy there is a decrease in the specific entropy of the material falling into the core which begins to contract once more.

Radiative energy transfer assumes more importance now and with the loss of quite significant amounts of energy (by both convective transfer from the interior and radiative

45

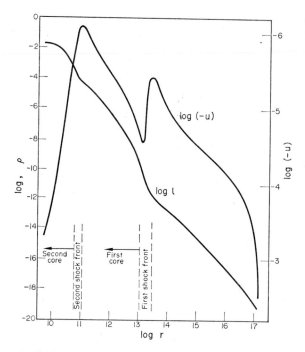

FIG. 17. Density and velocity distributions following the formation of the second core. Shock fronts are represented by the regions of steep positive slope in the velocity curve. (After Larson.)

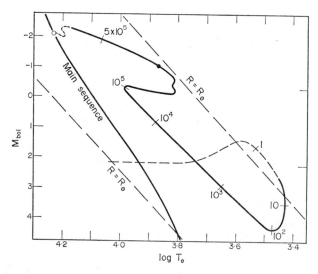

FIG. 18. The evolution of the stellar core for a protostar of 5 $M_\odot$. Initial temperature = 10°K. The dashed lines are isochrones marked in years. The time in years since formation of the core is marked along the curve. The solid dot represents the point where half of the mass has been accreted; the open circle where the star reaches the main sequence. (After Larson.)

losses from the surface), there is an appreciable contraction of the core. This contraction occurs between 10 and 100 years after the formation of the core (Fig. 18).

This particular evolutionary phase is analogous to the contraction of a pre-main sequence star along the Hayashi track on the Hertzsprung–Russell diagram. Owing to the presence of dust grains in the outer layers of the cloud, the energy transmitted is in the form of thermal infrared radiation from the dust grains.

ACCRETION PHASE

We may look upon the infalling material as having been in free fall from an infinite distance and therefore the velocity u_2 at the point where it enters the shock front is given by

$$\tfrac{1}{2}u_2^2 = \frac{GM}{R} \tag{3}$$

where M is the mass of the core and R is its radius. From equation (3) we readily see that as R decreases, the kinetic energy of the infalling material increases and it soon exceeds the energy outflow from the interior of the core.

Instead of decreasing, the specific entropy of the core begins to increase as a result of an increase in the surface temperature and the associated decrease in the energy loss from the core. Contraction of the core ceases, the outer convection zone gradually disappears, while the mass of the core increases appreciably.

At this point, evolution proceeds somewhat differently than for protostars with masses less than $\sim 2.5\ M_\odot$. Accretion of matter from the surrounding protostellar cloud continues but now radiative energy transport in the interior of the core assumes a greater importance.

After about 2.3×10^4 years following the initial formation of the core, the surface layers are in radiative equilibrium. The original mass of the protostellar cloud is now distributed as $3.7\ M_\odot$ in the core and $1.3\ M_\odot$ still remains in the surrounding cloud. The surface temperature of the core is approximately 7000°K.

FINAL EVOLUTIONARY STAGE

The surface temperature and the luminosity of the core now begin to decrease due to a corresponding decrease in the kinetic energy flux. This happens when approximately half of the total mass has been accreted by the core. Two points are worthy of note here.

For protostars with masses greater than $\sim 3\ M_\odot$, such as we are now discussing, the central core evolves by radiative cooling and contraction to become a main sequence star while there is still an appreciable amount of infalling material remaining in the surrounding cloud. It is therefore on the main sequence by the time it becomes visible. Prior to this, it is still obscured by the protostellar envelope still in existence. Second, the radius of the core never becomes large enough for a convective phase to exist at this stage. The "Hayashi" phase, therefore, does not appear to exist at all for these more massive stars.

Comparison with Previous Models

The dynamics of a collapsing protostar as suggested above by Larson (1969, 1972) differ in some important ways from those of previous workers, notably Hayashi (1961, 1962). Here, the collapse of the protostellar cloud has been assumed to be either entirely homologous or consisting of polytropic density distributions. As we have seen, one result of such

The Nebular Variables

a collapse is that such protostars begin their evolution with very large radii of the order of 60–150 $R_\odot$.

It may be significant, however, that for stars of $\leqslant 1\ M_\odot$, the end result of both theories is virtually identical. The properties of newly-formed stars and of the stellar cores when they reach the main sequence, according to Larson's models, are given in Table V. These may be taken as the properties of such stars when they first become visible.

TABLE V

(A) PROPERTIES OF NEWLY-FORMED STARS

$M/M_\odot$	$\log T_e$	$R/R_\odot$	$L/L_\odot$	t (years)
0.25	3.57	1.8	0.55	2.8×10^5
0.50	3.60	1.9	0.87	5.0×10^5
1.0	3.64	2.1	1.6	8.7×10^5
1.5	3.67	2.8	3.3	1.2×10^6
2.0	3.83	4.0	30.0	1.4×10^6

(B) PROPERTIES OF STELLAR CORES WHEN THEY REACH MAIN SEQUENCE

M_{total}	M_{core}	$\log T_e$	$R/R_\odot$	$L/L_\odot$	t (years)
3.0	3.0	4.17	2.2	200	1.4×10^6
5.0*	4.0	4.23	2.7	520	8.7×10^5

*Initial temperature taken as 20°K instead of 10°K.

The Infrared Hertzsprung–Russell Diagram

During much of its evolutionary phase a condensing, or core-accreting protostar will not be detectable in visible light and even during some of the later stages of evolution, when the opacity of the surrounding, infalling material is diminishing, it will be primarily an infrared object. Larson (1972) has plotted an infrared Hertzsprung–Russell diagram (Fig. 19) to illustrate the evolution of a protostar. Here $\log (L/L_\odot)$ is plotted against $\log \lambda_m$, where λ_m is defined as the wavelength (μ) of the peak infrared emission (per unit wavelength interval).

The family of curves shown in Fig. 19 are only approximations since both the shape of the infrared spectrum for these stars and the value of λ_m are sensitive to the infrared opacity of the surrounding dust grains which is, as yet, not accurately known. The isochrones shown also in the diagram indicate that the evolutionary scale increases from right to left.

Because of this increasing time scale, most of the protostars which have been observed will possess comparatively high apparent temperatures of between 500 and 1000°K and few, if any, can be expected with low temperatures of the order of 100°K. It is also significant that most of the infrared objects that have so far been discovered lie in the upper left part of the diagram. Again, this is to be expected since the probability of discovery increases with the brightness of the object. From the position of R Monocerotis on the diagram, we can interpret this object as a protostar with a total mass of $\sim 5\ M_\odot$. The effective temperature of this variable, as derived from Fig. 19, appears to be about 1.5×10^{4}°K.

How does this compare with observation for this star? Unfortunately, this is not an easy question to answer since, as Herbig (1968) has shown, R Monocerotis does not have a

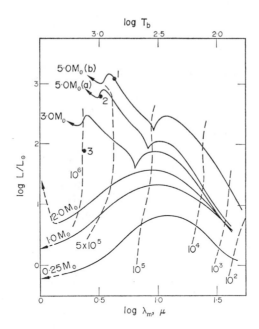

FIG. 19. Evolutionary tracks of protostars in an infrared Hertzsprung–Russell diagram. The dashed lines are isochrones marked in years. The infrared objects marked as solid dots are: 1, the Becklin–Neugebauer object; 2, R Monocerotis; 3, R Coronae Australis. (After Larson.)

normal stellar spectrum and it is possible that all of the observed spectral features may originate in the dense circumstellar shell. However, there are strong absorption lines of the Balmer series present in the spectrum which indicate a temperature for the central star of the order of 10^{4}°K.

Effect of Angular Momentum

It is, of course, unrealistic to expect that a star will form from a collapsing protostellar cloud with zero angular momentum. Various authors have discussed models of protostar formation which take the rotation of the initial cloud into consideration. Some of these are discussed later in Chapter 19 since they have been confined, more or less, to low mass stars which are more in keeping with what we expect the T Tauri variables to be. Here we shall examine those which appear to be applicable to stars of higher mass in the range $3\,M_{\odot}$ to $\sim 5\,M_{\odot}$.

The problem of how a star approaches the main sequence has been studied extensively in recent years and a good deal of progress has already been made in our understanding of these pre-main sequence stages.

Ezer and Cameron (1965) made the assumption that the material in the protostellar cloud begins with the highest possible adiabat that is consistent with the virial theorem which states that, for a given temperature, a low adiabat has a high density and vice versa.

Hayashi (1966) and Larson (1969, 1972) have studied the hydrodynamic collapse of a cloud to form a protostar assuming zero angular momentum and, although their approach to the problem differs regarding the nature of the collapse, both have shown that much of

49

the internal energy is radiated away so that the stellar material begins on an appreciably lower adiabat. The gas in the centre of such an object remains on a lower adiabat since it can be heated only by compression.

Not until the outer layers of the protostar fall upon the central core does shock heating (due to the release of gravitational potential energy) raise their adiabat. Once this occurs, the star is then able to form a stable body which is in hydrostatic equilibrium during the evolutionary phase prior to reaching the main sequence.

Under certain conditions it can be demonstrated that the collapsing protostellar cloud will form into the shape of a flattened disc due to rotation. Gaseous dissipation processes may then allow the formation of a star. At this stage, the gas density at the centre of such a disc will probably be similar to that in the core of Larson's models (Larson, 1969) at the point where the shock front is first formed.

The initial adiabat in the flattened disc prior to stellar formation has been estimated for three cases by Perri and Cameron (1973).

(a) For the onset of adiabatic compression when all of the hydrogen molecules exist in the parahydrogen form.

(b) For the onset of adiabatic compression when the parahydrogen to orthohydrogen ratio is 1:3 and

(c) An estimate which is based upon the temperature and pressure conditions calculated for the flattened protostellar disc at the time of meteorite accumulation.

As Perri and Cameron have pointed out, there is still some considerable uncertainty associated with the meteoritic cosmobarometers and cosmothermometers as shown by various authors including Jeffery and Anders (1970), Keays et al. (1971) and Laul et al. (1972). Nevertheless, it is possible to assume an approximate temperature of 450°K and a pressure of 5×10^{-6} atoms as the necessary conditions which define the adiabat.

Since, in all three cases, the adiabats are fairly low, it can be shown that the thermal dissociation of the hydrogen molecules is incomplete before any effects due to pressure dissociation become important. Extension of these adiabats in the direction of higher densities and temperatures has been carried out by Perri and Cameron (1973) allowing both for the internal energy of excitation of hydrogen molecules, and the dissociation of hydrogen and helium with subsequent ionization. The lowering of the dissociation energy at higher pressures was calculated by the method given by Vardya (1965).

Thermal ionization of hydrogen atoms is shown to be extremely inefficient, such that more than 50 per cent of these atoms remain neutral, at least until pressure ionization occurs. Stewart and Pyatt (1966) have shown that coulomb binding of electrons to the plasma will lower the ionization energy of hydrogen at higher densities, while pressure ionization, as indicated by Rouse (1967), is brought about mainly from the partial shielding of the electrons in the K-shell of neutral atoms by the free electrons lying within the classical Bohr radius.

After complete ionization, the adiabat may be written in the usual form of

$$P = K\rho^{5/3}. \tag{4}$$

Expressing both pressure and density in c.g.s. units, Perri and Cameron (1973) obtain the following values for the initial adiabat for the three cases considered: $K = 1.08 \times 10^{13}$, 1.67×10^{13} and 2.61×10^{13}.

The stellar model discussed by Perri and Cameron is assumed to have this final adiabat (the largest of the three) present throughout the whole of its interior structure. According to Chandrasekhar (1939) the star may then be represented by a polytrope of $n = 1.5$. This leads to a temperature of $7.0 \times 10^7 \, °K$ and a density of $7.2 \times 10^3 \, g \, cm^{-3}$ at the centre of the star.

An important outcome of these calculations is that, since the pressure in the centre of a rotating, flattened disc is virtually independent of the temperature, thus resembling an electron-degenerate core in this respect, hydrogen ignition in the core can result in a thermal runaway. The adiabat of the gas can then attain a higher value than would be reached on the main sequence, resulting in a body on the low temperature side of the main sequence, which is in hydrostatic equilibrium. Perri and Cameron suggest that the rise in luminosity during this stage of the hydrogen flash would be sufficiently rapid to give rise to an object such as FU Orionis.

This idea may also explain the anomalous situation found in several young clusters where many of the low luminosity stars are located very close to the low temperature boundary of the main sequence. Such a situation had earlier been taken by Iben and Talbot (1966) and Ezer and Cameron (1971) to imply that these low mass stars required an evolutionary time scale several millions of years longer than the associated high mass members. However, the dissipation time scales for flattened, rotating discs of a wide mass range are all comparable to within a few thousands of years which is considerably less than the dispersion in the interstellar collapse times. Consequently, the hydrogen flash process locates all of the stars in a young cluster, virtually regardless of their mass, in the same region close to the main sequence.

Toroidal Stages in Collapsing Protostellar Clouds

In most of the theories concerned with the collapse of a rotating protostellar cloud, prior to the formation of a stellar core (or cores), the only figure of equilibrium which is considered is the ellipsoid of revolution. Unfortunately, this often leads to results which are far from easy to explain. One consequence of this assumption is, of course, that the angular velocity increases considerably while the moment of inertia decreases during subsequent stellar evolution.

We also find that the angular momentum of a protostar exceeds that of main sequence stars by a factor as high as 10^4 and, although the presence of magnetic fields may be invoked to remove this difficulty we are then faced with the problem of explaining why the magnetic stars are rapid rotators when they should have lost most of their angular momentum. A further problem is why the stars with early spectral types show a general increase in rotational velocity with an increase in their radii.

Porfiriev *et al.* (1969) have suggested that most of these problems disappear if figures of equilibrium other than the ellipsoid of revolution are considered. Such alternative figures appear necessary if the rotation is non-uniform as will arise if the collapse is non-homogeneous. Surfaces of the fourth or higher orders may be either one- or two-binding depending upon their parameters. On the assumption that the surface of the gas and dust cloud is of this type, it can be demonstrated that the protostellar cloud will eventually collapse into a two-binding body, namely a toroid.

Lichtenstein (1922) first proved the existence of such an equilibrium form for a homogeneous incompressible liquid while the similar case of a thin polytropic toroid has been examined by Ostriker (1964).

The contraction of an initially rotating protostellar cloud into a two-binding body that then degenerates into a thin toroid has been examined by Porfiriev and his colleagues who have shown that during this process there is little change in the moment of inertia and consequently the rotational velocity remains low. Only the evolution of a sufficiently thin toroid is capable of treatment at the moment since there is, as yet, no exact solution to the problem of a gaseous toroid of an arbitrary thickness.

The contraction of a thin toroid will be essentially isothermal owing to the low density of the gas and dust cloud and numerical calculations indicate that there is only a small change in the major radius but the minor one diminishes by a factor of between 10^3 and 10^4 on a time scale of the order of 5×10^5 years.

Uniform contraction will persist until the configuration becomes gravitationally unstable. At this point, when the number of the Jeans waves at the length of the major circle takes on an integral value, fragmentation occurs. The minimum of the total potential energy of the system is approached when there is maximum symmetry and one result is that during the fragmentation of the toroid, stars having different masses will be produced. A further outcome of this theory is that it satisfactorily explains the origin of polar magnetic fields since these are formed naturally during the fragmentation process, due to pinching effects as the rotating stars are formed.

The individual momentum of a star according to this theory is of the order of $(a/R)^2$ where a and R are the minor and major radii respectively.

COMPARISON WITH OBSERVATION

The mass distribution of the fragments formed with a ring of $100\,M_\odot$ and excitation of the twentieth harmonic has been calculated by Porfiriev *et al.* (1969). This shows that the total number of stellar-like fragments will be 128 with the majority possessing masses between 0.25 and $1.0\,M_\odot$. Only about 4 per cent will have masses $>2.25\,M_\odot$.

Recently, Isserstedt and Schmidt-Kaler (1968) have discovered a very large number of objects termed "Sternringen" which appear to be rings of stars such as would be formed by the fragmentation of a toroid. Further observational data are still required, however, to prove conclusively that these objects are, indeed, gravitationally-bound rings of stars as required by this hypothesis.

References

AVENI, A. F. and HUNTER, J. H. JNR. (1970) *Mém. Soc. Roy. Sci. Liége* 19, 65.
BODENHEIMER, P. and SWEIGART, A. (1968) *Astrophys. J.* 152, 515.
BREGER, M. (1972) *Ibid.* 171, 539.
CHANDRASEKHAR, S. (1939) *Introduction to the Study of Stellar Structure*, Univ. of Chicago Press, Chicago.
EZER, D. and CAMERON, A. G. W. (1965) *Can. J. Phys.* 43, 1497.
EZER, D. and CAMERON, A. G. W. (1971) *Astron. and Space Sci.* 10, 52.
FIELD, G. B. (1970) *IAU Symposium No. 39: Interstellar Gas Dynamics*, p. 51, Reidel, Dordrecht.
GAUSTAD, J. E. (1963) *Astrophys. J.* 138, 1050.
HAYASHI, C. (1961) *Publ. astr. Soc. Japan* 13, 450.
HAYASHI, C., HOSHI, R. and SUGIMOTO, D. (1962) *Prog. Theor. Phys. Suppl.*, No. 22.
HAYASHI, C. and NAKANO, T. (1965) *Prog. Theor. Phys.* 34, 754.
HAYASHI, C. (1966) *A. Rev. Astr. Astrophys.* 4, 171.
HEILES, C. (1971) *Ibid.* 9, 293.
HERBIG, G. H. (1968) *Astrophys. J.* 152, 439.
IBEN, I. JNR. and TALBOT, R. J. (1966) *Ibid.* 144, 968.
ISSERSTEDT, I. and SCHMIDT-KALER, T. (1968) *Veröff. Astr. Inst. Ruhr Univ. Bochum*, No. 1.
JEFFERY, P. M. and ANDERS, E. (1970) *Geochim. Gosmochim. Acta* 34, 1175.
KEAYS, R. R., GANAPATHY, R. and ANDERS, E. (1971) *Ibid.* 35, 337.
LARSON, R. B. (1969) *Mon. Not. Roy. astr. Soc.* 145, 271.

LARSON, R. B. (1972) *Ibid.* **157**, 121.

LAUL, J. C., KEAYS, R. R., GANAPATHY, R. and ANDERS, E. (1972) *Geochim. Gosmochim. Acta* **36**, 329.

LICHTENSTEIN, L. (1922) *Math. Zeitschr.* **13**, 82.

McCREA, W. H. (1957) *Mon. Not. Roy. astr. Soc.* **117**, 562.

MENON, T. K. (1966) *Trans. IAU, Vol. XIIB*, 445, Academic Press, London and New York.

MOSS, D. L. (1973) *Mon. Not. Roy. astr. Soc.* **161**, 225.

OSTRIKER, J. (1964) *Astrophys. J.* **140**, 1067.

PENSTON, M. V. (1966) *Roy. Obs. Bull.* No. 117, 299.

PERRI, F. and CAMERON, A. G. W. (1973) *Nature* **242**, 395.

PORFIRIEV, V. V., SHULMAN, L. M. and ZHILYAEV, B. E. (1969) *Ibid.* **222**, 255.

ROUSE, C. A. (1967) *Phys. Rev.* **159**, 41.

STEWART, J. C. and PYATT, K. D. (1966) *Astrophys. J.* **144**, 1203.

STROM, K. M., STROM, S. E. and YOST, J. (1971) *Ibid.* **165**, 479.

STROM, S. E., STROM, K. M., BROOKE, A. L., BREGMAN, J. and YOST, J. (1972a) *Ibid.* **171**, 267.

STROM, S. E., STROM, K. M., YOST, J., CARRASCO, L. and GRASDALEN, G. (1972b) *Ibid.* **173**, 353.

VARDYA, M. S. (1965) *Mon. Not. Roy. astr. Soc.* **129**, 345.

WRIGHT, A. E. (1970) *Mém. Soc. Roy. Sci. Liége* **19**, No. 195.

PART II

T Orionis Variables

CHAPTER 8

Light variations of T Orionis stars

THE PROTOTYPE star of this group of the nebular variables is T Orionis discovered by Bond in 1863 and, incidentally, the first variable star to be discovered at the Harvard University Observatory. Wide variations in the amplitude of this star have been noticed over a long period of years, particularly in the visual, with the extreme range in V being from $9^m.5$ to $12^m.1$. For many years now, this variable has fluctuated just below its maximum brightness with an amplitude of $\sim 1^m$.

The star lies within the Orion Nebula, in the same field as the Trapezium (θ Orionis) and its light variations are well within the reach of even small instruments. Consequently, it is one of the more completely observed of all these variables.

From the light curve (Fig. 20) it will be seen that the variations in brightness are different from those of the majority of the RW Aurigae or T Tauri stars. The curve is generally characterized by non-periodic and abrupt minima together with slow and short fluctuations during periods of "normal" light.

Like almost all of the nebular variables, the T Orionis stars are very young objects associated with nebulosity. It is, perhaps, significant that the fluctuations found around maximum or median brightness seem to be characteristic of such stars. In certain of these variables, for example, SV Cephei, these are even more pronounced than in the case of T Orionis (Fig. 21).

Two further examples of this kind of behaviour are provided by WW Vulpeculae (Fig. 3) and SU Aurigae (Fig. 22). Both show the sharp minima associated with the T Orionis type of light curve, but whereas the spectra of T Orionis, SV Cephei and WW Vulpeculae are all of Type A, that of SU Aurigae is G2 and it is possible that we should classify the latter among the RW Aurigae variables.

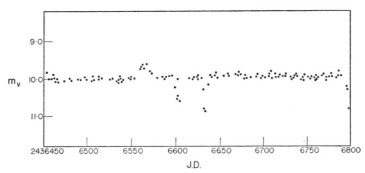

FIG. 20. Light curve of T Orionis.

The Nebular Variables

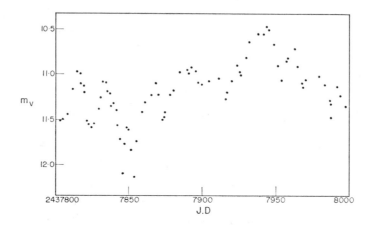

FIG. 21. Light curve of SV Cephei.

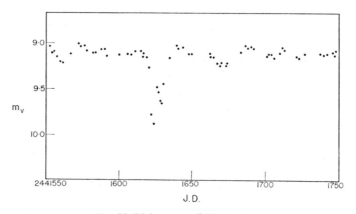

FIG. 22. Light curve of SU Aurigae.

The amplitudes of these stars, as may be seen from the data in Table VI, are comparable with those of the RW Aurigae variables and, in general, somewhat larger than those of the T Tauri stars.

Three-colour Photometry of T Orionis Variables

In addition to the very comprehensive visual and photographic light curves obtained for T Orionis by Parenago (1955) and for the closely related variable DD Serpentis by Meinunger (1967), three-colour photometry on the UBV system has been carried out on certain of these variables, notably the photoelectric observations made by Wenzel (1969).

The V/B–V and U–B/B–V diagrams for WW Vulpeculae shown in Figs. 23 and 24 illustrate the small scatter of the colour indices U–B and B–V for this star compared with the amplitude of the corresponding visual variations.

In the V/B–V diagram, the solid line represents the direction of the main sequence in the respective interval of B–V, while the arrow shown in the U–B/B–V diagram indicates the interstellar extinction which was not applied to the observations.

TABLE VI. T ORIONIS VARIABLES

Star	Magnitude		Spectrum
	Maximum	Minimum	
AB Aur	7.2	8.4	A0ep
HR Car	8.2	9.6	B:ep
SV Cep	10.1	12.1	A
R CrA	10.0	13.6	F5
S CrA	10.8*	12.5	G5
T CrA	11.7*	13.5	F0
TY CrA	8.7*	12.4	B2
XX Oph	9.1	10.7	B:ep
IX Oph	11.8	12.7	—
T Ori	9.5*	12.1	A2
TV Ori	13.4	15.6	K0
UX Ori	8.9	10.6	A2e
UY Ori	11.5	13.2	—
AN Ori	10.7*	11.7	K1e
BN Ori	9.3*	11.4	A7
X Per	6.0*	6.6	B0e
XY Per	8.8*	10.1	A2 II+B9
AT Sco	13.3	15.2	—
AU Sco	13.8	14.7	—
WW Vul	10.9	12.6	A2

*Visual magnitudes, all others being photographic.

This narrow scatter of the points in these diagrams is in marked contrast to the wide range of points found by Wenzel (1966) for RW Aurigae. It is also of interest to compare the V/B–V diagram of WW Vulpeculae with that of RY Tauri (Fig. 38).

Infrared Observations of T Orionis Variables

As mentioned in Chapter 2, the near infrared observations made by Mendoza (1966, 1968) suggested that much of the luminosity of these gravitationally contracting stars might lie in the infrared rather than in the visible spectrum. With the recent development of multifilter narrow-band photometers it is now possible to measure the intensity of stellar radiation as far as 22μ within a single night, greatly extending our information of the energy distribution of these stars.

Cohen (1973a) has carried out multifilter observations of several T Orionis variables lying in the Orion complex and the Perseus–Auriga–Taurus dark clouds. These may be conveniently divided into two groups.

(a) Irregular variables having early-type spectra, mostly without emission lines, typified by variables such as XY Persei and CQ Tauri (Cohen, 1973a).

(b) Early-type stars which appear nebulous, for example, AB Aurigae and V380 Orionis (Cohen, 1973b).

Figure 25, taken from Cohen's observations, shows certain interesting features. The energy distribution curves are all noticeably flat over the entire spectral range, similar to those

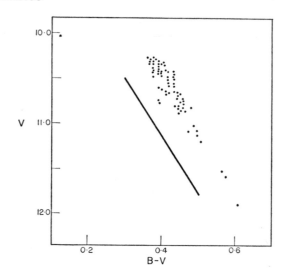

FIG. 23. V/B–V diagram for WW Vulpeculae. (After Wenzel.)

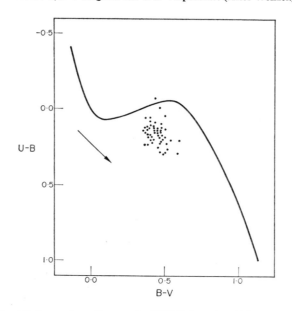

FIG. 24. Two-colour diagram for WW Vulpeculae. (After Wenzel.)

of CO Orionis (Fig. 13) and those of the T Tauri variables T and RY Tauri (Fig. 40). There is a notable feature in the region of 10μ for XY Persei which was found to be even more marked in the case of AB Aurigae and V380 Orionis.

That these features are genuine, particularly for the latter two variables, was shown by carrying out similar observations on different nights using other multifilter systems and detector units but with identical filters. Earlier observations made by Gillett and Stein (1971) are also in good agreement with these results over this spectral range. It is also perhaps significant that such features have been observed in the infrared spectrum of Comet Bennett by Maas *et al.* (1970). The latter have attributed these features to the presence of silicate grains. Cohen (1973b) has shown that the energy distribution curves for AB Aurigae and V380 Orionis over this spectral range are very like that for Comet Bennett.

As for the RW Aurigae variables, the marked flatness of the curves is in agreement with the hypothesis that here we have a circumstellar shell (or ring) of solid material surrounding these stars. The fact that such stars are probably rotating rapidly would suggest that this cool material has a disc-like distribution situated equatorially around the star.

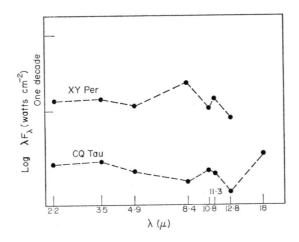

FIG. 25. Infrared energy distribution for two typical T Orionis variables. (After Cohen.)

Orion Flare Stars

Here we shall make a distinction between the classical flare stars of the UV Ceti type and those which may be termed the Orion flare stars.

The former are dMe-type red dwarfs with small masses between 0.04 $M_\odot$ (UV Ceti) and 0.16 $M_\odot$ (Do Cephei) as determined by Petit (1961). All are situated in the solar neighbourhood where there are no known T Tauri variables. Accordingly we shall not discuss these flare stars further, confining our attention to the numerous flare stars found in the Orion Nebula and the young T-association near NGC 7023. These not only show a spatial distribution similar to the T Tauri and T Orionis variables but may, as suggested by Haro and Chavira (1965), represent the evolutionary stage following the T Tauri stage.

Over the past twenty years, a wealth of data have accumulated on flare stars in stellar aggregates having different ages due, mainly, to the efforts of Haro and Chavira (1968) at Tonantzintla and Rosino (1969) at Asiago. The majority of these discoveries have been made in the region of the Orion Nebula. Like the numerous T Orionis and T Tauri variables, these flare stars exhibit a strong concentration towards the centre of the Orion Nebula and, as shown by Kholopov (1959) are almost certainly members of a T-association (Orion T2).

Two fields in the Orion Nebula have been extensively covered by Rosino (1969); the region around the Trapezium which includes the four nebulae NGC 1976, 1977, 1982 and 1999 and the area around ζ Orionis covering both the Horsehead Nebula and NGC 2024. Rosino (1946, 1956, 1962) has shown that the density of variable and flare stars around the Trapezium is particularly high, approximately 70 per cent of the stars in this region being

The Nebular Variables

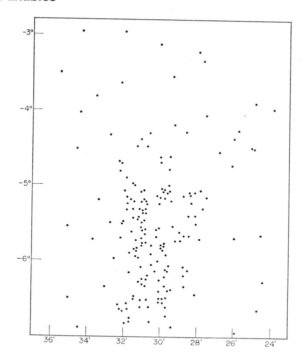

FIG. 26. Distribution of flare stars in the Orion aggregate. (After Rosino.)

variable. Another area containing a very high number of T Orionis (and T Tauri) variables is in a lane passing from NGC 1977, through NGC 1999 and on in the direction of the Horsehead Nebula. Rosino (1962) also demonstrated that whereas variables occur in heavily obscured regions, they are mostly to be found on the fringes of dark nebulosities.

The very close similarity between the distribution of Orion flare stars and nebular variables in the Orion complex is brought out in Figs. 26 and 27.

Although this correlation is obvious, the dependence is not as strong as was originally thought, but it is sufficiently so to suggest that the flare stars in the Orion Nebula are members of the T-association.

All of the flare stars discovered in these two regions by Rosino are fainter than photographic magnitude $16^m.5$ and have a mean value of $17^m.2$. From the observations made at Asiago it appears that, although the majority of these flare stars in the Orion Nebula lie to the right of the main sequence, a small number lie on it and a few are found to the left.

The results also indicate that the mean interval between successive flares is greater than 10 days and statistically, it seems that a flare on these stars is a comparatively rare event. This is in general agreement with the observation of Haro and Chavira (1968) that the frequency of flares in a given star depends both upon the nature of the stellar aggregate and its age. While typical flares are generally found in stars which are essentially constant in brightness outside the time of a flare, a few of the closely associated nebular variables have been observed to flare. In this case, however, the amplitude of the flare is smaller, of the order of $1^m.1$, than in the flare stars themselves (amplitude $= 2^m.0$).

In this context we must draw a distinction between actual flares observed in nebular variables and the rapid fluctuations of brightness that are sometimes observed in T Orionis

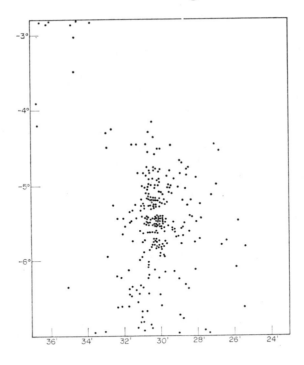

FIG. 27. Distribution of nebular variables in the Orion aggregate. (After Rosino.)

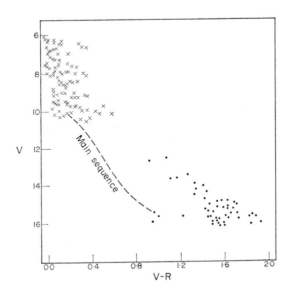

FIG. 28. Colour magnitude diagram for Orion flare stars (filled circles). Crosses represent bright members of the Orion association. (After Andrews.)

The Nebular Variables

variables (typical stars being SY, XX, YY and YZ Orionis). Although these variations have amplitudes of 1^m or more, they are not generally regarded as flares.

Multicolour photometry of the Orion flare stars has been carried out by Andrews (1969) utilizing material obtained with the Boyden 60-in. reflector and a magnetically-shielded photomultiplier cooled to 0°C. Earlier multicolour photometry for seven of these stars has been made by Mendoza (1968) who was able to show that they possess large infrared colour excesses similar to the T Tauri variables.

While there is admittedly a considerable scatter of the Orion flare stars about the main sequence in the V/B–V diagram, Andrews has shown that this is considerably reduced in the corresponding V/V–R diagram (Fig. 28) where these stars fall in a band about 1^m to 2^m above the main sequence.

From Fig. 28 it will also be apparent that there is a clear-cut leftward limit at V–R $\sim 1^m.0$, this being defined by the brighter stars of this group which are of spectral type K0 to K1. Comparison with the classical UV Ceti flare stars shows that in the B–V/V–R diagram these lie within, but to the red of, the Orion flare star region and, with the exception of DH Carinae, they fall below the reddest of the Orion stars in the U–B/B–V diagram.

Flare Stars Near NGC 7023

A small T-association near the nebula NGC 7023 has been examined for flare stars similar to those found in the Orion Nebula by Mirzoyan and Parsamian (1969). This particular T-association is of interest because not only is it very compact but it is also comparatively close to us, at a distance of only 280 pc according to Weston (1953). Nine flare stars have been discovered in this region by Mirzoyan et al. (1968) and a tenth possible candidate by Rosino and Romano (1962).

The limiting absolute magnitude for the plates taken by Mirzoyan and Parsamian (1969) was $+11^m$ and, assuming that the probability of a flare P_k obeys Poisson's law, the following equations may be used to estimate the total number of flare stars in this T-association.

$$P_k = \frac{e^{-vt}(vt)^k}{k!} \tag{1}$$

$$N_k = NP_k \tag{2}$$

where t is the period of observation, v the frequency of flares, k is the number of flares during time t and N_k is the number of stars which have flared k times during the observation period.

From the data obtained by Mirzoyan and Parsamian, the total number of flare stars in this region is 27, while the average interval between flares is $3^d.3$.

The distribution of the flare stars so far discovered near NGC 7023, shown in Fig. 29, is of particular interest. It shows that eight of the stars lie some distance from the centre of the T-association, possibly beyond the limits as indicated by the circle. The suggestion that the flare stars actually surround the association and the two which appear within it are almost certainly projected upon it, is probably correct. It appears unlikely that these are field stars since this would imply a very high frequency of flares among the field stars which has not been observed.

On the assumption that Haro's hypothesis is correct and the flare stage immediately follows the T Tauri stage in these very young objects, their distribution around NGC 7023 may be interpreted as evidence that the T-association is expanding. As Mirzoyan and

64

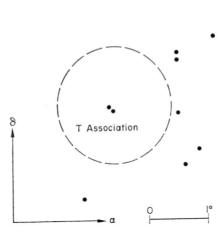

Fɪɢ. 29. Distribution of flare stars near NGC 7023. (After Mirzoyan and Parsamian.)

Parsamian have pointed out, however, the present statistical sample is not large enough for a final conclusion on this point to be reached.

Flare Stars in the Taurus Dark Clouds

To date, eleven flare stars have been discovered in the Taurus dark clouds. The spectral types of these variables are all between M0 and M5 and, in general, none of them show the Hα emission line during their quiescent periods. The amplitudes are smaller than the major flares of typical UV Ceti variables, being between $0^m.5$ and $1^m.7$ photographic.

None of the eleven stars in this region has been observed to flare more than once and the number of flares per hour of observing time (0.05) is considerably lower than the corresponding figures for similar stars in the Orion Nebula (0.37) and in the region of NGC 2264 (0.24).

The Flare Star Evolutionary Stage

From the general characteristics of the flare stars which have been discovered in various stellar aggregates, particularly young clusters, it appears that, following Haro and Chavira (1965) flare activity begins just after the T Tauri stage once the star has reached the main sequence. During the first stages of this evolutionary phase, the stars have earlier spectral classes than the members of the final UV Ceti stage and their luminosities are higher. The overall development of this phase corresponds very closely to the increasing age of the stellar aggregates in which the stars are found. Not only this, but there is a similar trend in the intensities of the emission lines, these being weak for young stars and becoming increasingly strong with stellar age, reaching their greatest intensity in the UV Ceti variables found in the solar neighbourhood.

Naturally, the spectroscopic behaviour of the emission lines during the flares of stars in the very young clusters (the Orion Nebula and the Taurus dark clouds, for example) has not been investigated to the same degree as those of the classical UV Ceti type. However, if we assume that this behaviour is similar in both types, we can draw some general conclusions regarding the energy source for this flare-like activity.

The Nebular Variables

The changes in the emission lines of both solar and UV Ceti type flares is very similar. Both types have several physical parameters in common and are also accompanied by radio emission. At maximum brightness during a flare there are very few absorption lines visible in the spectrum. Very intense emission lines of the Balmer series and Ca II are present with somewhat less intense lines of He I and He II. The underlying continuum shows a considerable strengthening in the violet region and this, taken in conjunction with the variations in the emission lines, suggests that the temperature rises from ~ 3000 to $\sim 10,000°K$ during a flare. Indeed, during a major flare of YZ Canis Minoris, the total energy radiated has been computed by Kunkel (1970) to be 6.6×10^{34} erg. The effective temperature on this occasion was $T_e = 20,000°K$.

From the analogy with solar flares, the energy source appears to be a magnetic field and the associated magnetohydrodynamic phenomena brought about by convective motions and by rotation. Poveda (1964) has theorized that the magnetic field may be localized in certain regions and is also periodic in nature, thereby being able to both regenerate the field and prevent its decay.

In particular connection with the flare stars found among the T-associations, the hypothesis elaborated by Gershberg (1970) and Kunkel (1967) may be more appropriate. Here it is assumed that well-defined regions of high temperature on the surface of a cool star heat a surrounding hot envelope. While this idea has much to commend it, there is, nevertheless, the difficulty of explaining the very rapid rate of formation of these flares, especially those of the Orion type flare stars. Here, the duration of a typical flare, as determined with high resolving in time by Christaldi and Rodono (1971a, 1971b), can be less than 15–20 sec.

References

ANDREWS, A. D. (1969) *Non-Periodic Phenomena in Variable Stars*, p. 137, Reidel, Dordrecht.
CHRISTALDI, S. and RODONO, M. (1971a) *Int. Bull. Variable Stars*, No. 525.
CHRISTALDI, S. and RODONO, M. (1971b) *Ibid.* No. 526.
COHEN, M. (1973a) *Mon. Not. Roy. astr. Soc.* **161**, 97.
COHEN, M. (1973b) *Ibid.* **161**, 105.
GERSHBERG, R. E. (1970) *Flares of Red Dwarf Stars*, Moscow.
GILLETT, F. C. and STEIN, W. A. (1971) *Astrophys. J.* **164**, 77.
HARO, G. and CHAVIRA, E. (1965) *Vistas in Astronomy* **8**, 89.
HARO, G. and CHAVIRA, E. (1968) *Stars and Stellar Systems* **7**, 141, Univ. Chicago Press.
KHOLOPOV, P. N. (1959) *Sov. astr. A. J.* **3**, 291.
KUNKEL, W. E. (1967) *An Optical Study of Stellar Flares*, Univ. Texas, Austin.
KUNKEL, W. E. (1970) *Astrophys. J.* **161**, 503.
MAAS, R. W., NEY, E. P. and WOOLF, N. J. (1970) *Ibid.* **160**, L101.
MEINUNGER, L. (1967) *Mitt. veränderl. Sterne* **3**, 20.
MENDOZA, E. E. (1966) *Astrophys. J.* **143**, 1010.
MENDOZA, E. E. (1968) *Ibid.* **151**, 977.
MIRZOYAN, L. V., PARSAMIAN, E. S. and CHAVUSHIAN, O. S. (1968) *Soobshch. byurak Obs.* **39**, 3.
MIRZOYAN, L. V., and PARSAMIAN, E. S. (1969) *Non-Periodic Phenomena in Variable Stars*, p. 165, Reidel, Dordrecht.
PARENAGO, P. P. (1955) *Trudy gos. astr. Inst. Sternberga* **25**, 216.
PETIT, M. (1961) *J. Observateurs* **44**, 11.
POVEDA, A. (1964) *Nature* **202**, 1319.
ROSINO, L. (1946) *Publ. Bologna* V, No. 1.
ROSINO, L. (1956) *Mem. Soc. astr. Ital.* **27**, 3.
ROSINO, L. (1962) *Ibid.* **32**, 4.
ROSINO, L. (1969) *Non-Periodic Phenomena in Variable Stars*, p. 173, Reidel, Dordrecht.
ROSINO, L. and ROMANO, G. (1962) *Asiago Contrib.* No. 127.
WENZEL, W. (1966) *Mitt. veränderl. Sterne* **4**, No. 4.
WENZEL, W. (1969) *Non-Periodic Phenomena in Variable Stars*, p. 61, Reidel, Dordrecht.
WESTON, E. B. (1953) *Astron. J.* **58**, 48.

CHAPTER 9

Spectroscopic characteristics

TAKEN as a whole, the spectroscopic characteristics of the T Orionis variables resemble those of the preceding class rather than those of the T Tauri variables to be discussed later. The spectra cover a fairly large range of early types and possibly include some members with late type spectra such as AA Orionis (spectrum K3) and AZ Orionis (spectrum K4). We may therefore say, that the spectral classes range from B to K with all intermediate types being represented, although Cohen (1973) restricts these variables to those irregular stars having only early type spectra, mostly without emission lines.

As mentioned earlier, one of the major characteristics looked for in an objective prism search for nebular variables is the presence of emission lines, particularly that of Hα. However, like the RW Aurigae variables, we may divide the T Orionis stars into two spectroscopic groups, those which show emission lines and those in which they are absent.

Hydrogen emission T Orionis Variables

Examination of the spectra of those T Orionis variables which show the characteristic emission lines, especially of Hα, indicates that there is no correlation between spectral type and emission lines. For example (Table VII), we find the Balmer lines in emission in stars with both early and late type spectra.

TABLE VII. HYDROGEN EMISSION T ORIONIS STARS

Star	Spectrum
T Ori	A3e
UX Ori	A2e
AN Ori	K1 IVe
AZ Ori	K4 IV–Ve

A study of the light variations of these stars also fails to show any relation between early or late spectral type and any noticeable characteristic of the light curves.

Non-emission T Orionis Stars

A large number of variables are found, particularly in the Orion complex and the Taurus–Auriga–Perseus dark clouds, which clearly belong to the T Orionis variables on the basis of their light curves, but which have spectra in which emission lines are absent. Quite possibly these represent stars which have evolved further than those members exhibiting emission lines although this is still a debatable point.

The Nebular Variables

TABLE VIII. NON-EMISSION T ORIONIS STARS

Star	Spectrum
R CrA	F5
TY CrA	B2
KS Ori	A0
MR Ori	A2 V

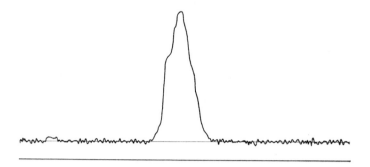

FIG. 30. Hα-line profile for V380 Orionis. (After Dibaj and Esipov.)

The Emission Spectrum

In those T Orionis stars having emission spectra, there is often a very close resemblance to the spectrum of RW Aurigae. This is particularly noticeable in the case of S Coronae Australis which is exceptionally rich in emission lines due to permitted transitions of ionized metals, e.g. Na I and Fe II, together with the H and K lines of Ca I and the multiplet of Ti II. The lines of Fe I at 4063 and 4132 Å are also present, as are those of He I. Weaker lines of Mg I and other metals also show in these spectra.

The Balmer series out to at least H19 and several lines of the Paschen series are a common feature of these variables. The rather broad, symmetrical profile of the Hα line obtained by Dibaj and Esipov (1969) shown in Fig. 30 is indicative of a rotating envelope around V380 Orionis which is a star seen "pole-on". This may be compared with Figs. 15, 46 and 47 which are intermediate cases between such stars as V380 Orionis and those having expanding envelopes with associated self-absorption such as RY Tauri and R Monocerotis.

Emission lines due to forbidden transitions are also present although somewhat weaker than the permitted lines. The presence and intensity of these lines vary from star to star. Those of [O I] at 6300 and 6363 Å are the only ones observed by Bonsack (1961) in the spectrum of S Coronae Australis, whereas AN Orionis shows only those of [S II] and [N II]. Forbidden lines of all three elements are present in the spectra of KR and LL Orionis. In the case of the emission lines of [N II], it would appear that these originate, not close to the stellar surface in these variables, but are emitted by the surrounding nebulosity.

The Absorption Spectrum

The intensity of the absorption lines in the spectra of the T Orionis stars varies considerably from one variable to another. In those stars where emission lines are absent, the absorption spectrum is often sufficiently well developed to enable an accurate estimate of the spectral type to be made. The absorption lines of several metals are found, particularly in those stars with intermediate or late type spectra (e.g. R Coronae Australis, F5 and TV Orionis, K0).

The absorption spectrum in those variables having prominent emission lines, varies considerably in strength, due mainly to line broadening and partial obscuration by the emission features. S Coronae Borealis exhibits no absorption lines at all. LL Orionis is peculiar among the stars of this group in having a very well-developed absorption spectrum. As shown by Bonsack (1961), however, this star is peculiar in that the Na I D lines are not visible at all, either in absorption or emission. A suggested explanation of this effect is that the sodium absorption and emission lines are undisplaced in this star and cancel each other out. Further observational data on this variable are required, however, before it can be decided whether this is indeed the case, or whether there exists a real deficiency of sodium.

Lithium Abundance in T Orionis Variables

The strong absorption doublet of Li I at 6708 Å has been observed in several T Orionis variables by Bonsack (1961) who measured the depth of the lithium lines and compared this with the mean depths of nearby absorption features of a similar intensity due, mainly, to Fe I. Fortunately, the lines of Li I and Fe I have virtually the same temperature sensitivity and this ratio of the depths of the lines provides a good measure of the lithium abundance in these stars. The ratios found are: AN Orionis, 2.5; KR Orionis, 2.7 and LL Orionis, 1.9.

Owing to the absence of an absorption spectrum in S Coronae Australis, it is impossible to determine any value for this ratio in this particular star. The figures just given correspond to a lithium/heavy metals ratio of ~ 400 times greater than that found in the Sun.

T Orionis Variables with Be Spectra

In general, these variables belong to the small subgroup of bright stars whose prototype is XX Ophiuchi. Typical members are HR Carinae which has been described by Hoffleit (1939), XX Ophiuchi described by Prager (1940) and X Persei. The first two have spectra of type B:ep and the last of B0e.

These spectra are characterized by a very large line broadening which may be due to either rapid rotation or turbulent motions in the atmospheres of these stars. Since both mechanisms affect the spectral lines in the same way it is extremely difficult to assess the relative importance of either mechanism in these stars.

Struve (1931) was the first to put forward the idea that the emission lines arise in gaseous rings ejected from rapidly rotating stars which may be at the limit of instability and this interpretation appears to be valid today.

Apart from the spectroscopic variations associated with the short-term fluctuations in brightness, there appear to be three other kinds of variations similar to those observed by McLaughlin (1961).

(a) An absorption spectrum due to either a ring or circumstellar shell is sometimes visible in addition to the emission lines. Variations in the intensity of the absorption lines occur due to changes in the density of such shells.

(b) The ratio of the intensity of the emission lines to the neighbouring continuum often varies (E/C variation). Such changes seem to be associated with the appearance and

The Nebular Variables

disappearance of the absorption spectrum of the shell. Generally, these variations are slow with a time scale of several years.

(c) At times, a narrow central absorption may be seen which divides the broader emission lines into a red and violet component. This is the case when the star is oriented such that the shell is projected against, at least, a portion of the photospheric disc. Frequently, these two emission components show variations in their relative intensities (V/R variations). These variations may be quasi-periodic on a time scale of several years.

Certain Be stars also show much shorter variations in V/R, e.g. Cassiopeiae which has been examined by Hutchings (1967) using photoelectric scanning techniques. So far, however, this technique has not been applied to the XX Ophiuchi variables. The existence of such rapid and irregular V/R variations again suggests rapid turbulent motions in the atmospheres of these stars.

It is, perhaps, significant that TY Coronae Australis, which does not belong to the XX Ophiuchi subgroup, does not show convincing evidence of emission lines in its B2 type spectrum.

T Orionis Variables with Early Type Spectra

Apart from the variables with B type spectra mentioned above, a large number of these stars have early type spectra of classes A and F, with and without emission lines. Some of these are giant stars as shown by their luminosity class, e.g. XY Persei (spectral types A2 II+B9).

In certain of the variables having emission spectra, the Hα line is particularly intense, e.g. UX Orionis (Lause, 1933).

T Orionis Variables with Later Type Spectra

A second large concentration of these variables have spectra of type K and, again, emission lines are present in some cases and absent in others. The majority of these stars, however, do not show emission lines, e.g. TU Orionis (K0), TV Orionis (K0), AH Orionis (K:V) and IU Orionis (K2III).

Judging from the luminosity classes which are available for those stars with later type spectra, they all belong to the dwarf class.

Very few, if any, of the variables belonging to the T Orionis type have spectra later than type K, thus differentiating them from the closely allied T Tauri variables.

References

BONSACK, W. K. (1961) *Astrophys. J.* **133**, 340.
COHEN, M. (1973) *Mon. Not. Roy. astr. Soc.* **161**, 97.
DIBAJ, E. A. and ESIPOV, V. F. (1969) *Non-Periodic Phenomena in Variable Stars*, Reidel, Dordrecht.
HOFFLEIT, D. (1939) *Harvard Bull.* No. 913.
HUTCHINGS, J. B. (1967) *Observatory* **87**, 289.
LAUSE, F. (1933) *A.N.* **250**, 82.
MCLAUGHLIN, D. B. (1961) *J.R. astr. Soc. Can.* **55**, 73.
PRAGER, R. (1940) *Harvard Bull.* No. 912.
STRUVE, O. (1931) *Astrophys. J.* **73**, 94.

CHAPTER 10

Physical characteristics of T Orionis stars

SOME confusion has existed in the past over the classification of these variables. In the first edition of the General Catalogue of Variable Stars by Kukarkin and Parenago (1948) they were classed as "Orion-type variables", whereas in the later edition by Kukarkin *et al.* (1958) they were generally isolated by the "RW Aurigae" designation. Cohen (1973) has defined them more rigidly as irregular variables found in nebulous regions and having early spectral types, mostly without emission lines. This latter classification is clearly in good agreement with several of the variables listed in Table VI.

Nevertheless, it is obvious that if we include such variables as S Coronae Australis, TV and AN Orionis, for on the basis of their light variations they clearly belong to this group, it is necessary to extend the range of spectral types quite considerably. This, in turn, implies that once more we are dealing with a heterogeneous class of objects covering a wide range of masses, effective temperatures and luminosities. The XX Ophiuchi variables which seem to form a subgroup of the T Orionis stars, also possess certain characteristics which are different from the main class.

When we consider the probable evolution of these variables and the possible variations in the initial and boundary conditions for the protostellar clouds from which they form, this is perhaps to be expected.

Masses of the T Orionis Variables

Direct determination of the masses of the T Orionis variables is beset with the same difficulties as we found for the RW Aurigae stars. Almost certainly they still have the remnants of dusty circumstellar shells around them of varying opacities. Evidence for this comes from the observations of Strom, Strom and Yost (1971) and Strom *et al.* (1972a, 1972b).

According to Knacke and his colleagues (1973), the masses of several early type variables in the region around R Coronae Australis lie between 1 and 4 $M_\odot$. This agrees very well with the theoretical picture of R Coronae Australis itself computed by Larson (1972) of a protostar with a total mass of $\sim 2.5\,M_\odot$ in which virtually all of the mass has been accreted by the star. A similar mass appears probable for V380 Orionis.

From all of these determinations, both observational and theoretical, it would appear that (a) the masses of the T Orionis variables cover a fairly narrow range and (b) they are intermediate between those of the RW Aurigae variables on the one hand and the T Tauri stars on the other.

Effective Temperatures of the T Orionis Variables

The T Orionis variables tend to fall into two groups as defined by their spectral types.

The Nebular Variables

A large number have spectra of types B and A and effective temperatures in the range 2.5×10^4 to $1.1 \times 10^{4°}$K. An approximately equal number have later type spectra around type K, particularly the group of such variables in Corona Australis and a small association near ϵ Orionis. The latter has been discussed by Shapley (1924) and, although their amplitudes are fairly small (between $1^m.2$ and $1^m.8$ photographic), their light variations are of the T Orionis type with non-periodic minima and small fluctuations around "normal" light. The effective temperatures of these variables are of the order of $8.0 \times 10^{3°}$K.

THE CIRCUMSTELLAR GAS AND DUST CLOUDS

Much of the infrared radiation from these variables appears to be produced by thermal re-radiation from the surrounding shells and, from the properties of this infrared emission, these have effective temperatures, like the shells around the RW Aurigae stars, of between 600 and 900°K. The fact that, in general, these shells appear to be thinner and contain less mass than those of the RW Aurigae and T Tauri stars may be accounted for by (a) more material having been accreted by the central star and (b) more of the material having been dispersed from the vicinity of the newly-formed star by the action of radiation pressure, particularly in the case of the high-temperature group of variables.

Luminosities of the T Orionis Variables

The luminosities of the early type T Orionis variables seem to be somewhat higher than those of most other nebular variables, ranging from 70 to 200 $L_\odot$. The second group with intermediate to late type spectra, are appreciably less luminous (approximately 20–40 $L_\odot$). A significant number of the early type variables in this class are probably giants or subgiants rather than typical dwarfs.

Stellar Radii

We may obtain some indication of the radii of these variables from their luminosities and spectral types (with due caution, however, since their spectra are, as we have seen, far from normal). The majority of the T Orionis variables do not appear to exhibit the usual dwarf characteristics in their spectra such as weak enhanced lines and relatively strong "low-temperature" lines. Representative radii appear to be between 5 and 15 $R_\odot$.

The theoretical calculations made by Larson (1972) yield somewhat lower figures for these values for a stellar core of mass $\sim 3.5\ M_\odot$ by the time it reaches the main sequence, namely $\sim 2.5\ R_\odot$. The corresponding figure based upon the homologous collapse of the protostellar cloud obtained by Ezer and Cameron (1965) is $\sim 80\ R_\odot$.

Binary Systems among the T Orionis Variables

Binary systems have been discovered among all these nebular variables of the three main classes. Their existence follows, almost inevitably, from most of the theories of star formation which take rotation of the protostellar cloud into account. Fragmentation of the unstable ring structure which is a common feature of these hypotheses leads to binary or multiple systems and, since the former represents the most stable configuration, star ejection from multiple systems can lead to the eventual formation of binaries even among these systems.

S CORONAE AUSTRALIS

This variable has been fully discussed by Gaposchkin (1937). The separation of the two components is 1″ and, so far as is known at present, only the brighter of the two is variable. The spectral type of the fainter component (photographic magnitude = 13.5) is not known.

THE ORION FLARE STARS

The close association of flare stars in certain regions (e.g. the Orion Nebula) with aggregates of typical T Orionis and other nebular variables was pointed out in Chapter 8. This has been taken to indicate that these flare stars are also members of the T-associations and provides some confirmation of the hypothesis put forward by Haro and Chavira (1965) that the flare stage is that which immediately succeeds the T Tauri phase.

Here we shall examine the relation between these flare stars and the classical UV Ceti variables and thus the probable number of binaries among the former. From the multi-colour photometry carried out by Andrews (1969) on the Orion flare stars, there is little doubt that these variables fall in the B–V/V–R diagram in a region which overlaps that of the UV Ceti stars, although the latter lie a little to the red of this band. In the U–B/B–V diagram, the UV Ceti variables fall somewhat below the reddest of the Orion flare stars but here the picture is complicated by the very large ultraviolet excesses in the majority of the Orion variables. It is, therefore, difficult to determine a precise U–B colour for these flare stars.

Rosino (1969) has demonstrated that the frequencies of flares in stars of the Orion aggregate is lower than for the UV Ceti variables. This, however, is to be expected since Haro and Chavira (1965) have shown that the frequency of flares depends upon the nature of the stellar aggregate and increases with its age.

It thus appears that any differences between these two types of flare stars are differences of degree and not of kind.

The faintness of the Orion type flare stars, coupled with the dense nebulosity in this region (and in other T-associations), make it difficult to obtain spectra of sufficiently high dispersion to search for evidence of duplicity. The UV Ceti variables, on the other hand, are comparatively close to the Sun enabling good quality spectrograms to be secured comparatively easily.

Of the 25 UV Ceti variables that have been extensively observed spectroscopically, and photographically, 19 are known binaries. Of these, three are spectroscopic binaries, two are separated by < 1″ and the remainder are visual binaries. Among the visual binary systems, it is the fainter component which exhibits flare activity.

Although it is unwise to extrapolate, there is certainly circumstantial evidence that a large number of the Orion flare stars are binaries and, if they are indeed a later stage in the evolution of typical nebular variables, a similar number of the T Orionis stars are also double systems.

The XX Ophiuchi Variables

This small subgroup of the T Orionis variables is composed of intrinsically bright Be or Ae stars. From their light curves, e.g. SV Cephei (Fig. 21) and WW Vulpeculae (Fig. 3), it is evident that they exhibit the typical variations of the T Orionis variables, the characteristically sharp, non-periodic minima being very pronounced.

The Nebular Variables

The majority of these variables are apparently bright objects as may be seen from the data in Table VI. A small number of faint members are, however known, e.g. V540 Centauri ($14^m.8$–$16^m.2$ photographic) and TU Octantis ($13^m.6$–$16^m.4$ photographic). Whether this variation is an intrinsic property or a distance factor has not yet been determined, although the latter appears more probable.

X PERSEI AND THE X-RAY SOURCE 2ASE 0352+30

The possible identification of X Persei with the X-ray source 2ASE 0352+30 has been suggested by van den Bergh (1972). The optical spectrum of this star has been discussed by Crompton and Hutchings (1972). The lines are generally abnormally wide and diffuse with the emission lines of the Balmer series being prominent.

An attempt by Braes and Miley (1972) to detect radio emission at a wavelength of 21 cm failed to provide any positive results while a similar attempt by Baud and Tinbergen (1972) to observe circular polarization in visible light yielded no component greater than 0.02 per cent of the total intensity.

Using a polarimeter with a birefringence modulator of the kind described by Kemp (1969) with the Battelle 31-in. reflector, Stokes *et al.* (1973) have observed circular polarization of this star of the same order as Baud and Tinbergen, viz. 0.01 per cent of the total intensity in B and V filters and ~ 0.005 per cent in the red.

The observed polarization may be an intrinsic quality of the star itself or due to the circumstellar shell. Polarization due to processes occurring in the interstellar medium appears to be ruled out since these, as shown by Kemp and Wolstencroft (1972), would give a minimum signal in the V filter and signals of opposite sign in the B and R filters. In all three filters, Stokes *et al.* found the rotation of the electric vector to be in the same direction.

Further observational data are required before it can be established whether any of the other XX Ophiuchi variables are associated with possible X-ray source.

References

ANDREWS, A. D. (1969) *Non-Periodic Phenomena in Variable Stars*, p. 137, Reidel, Dordrecht.
BAUD, B. and TINBERGEN, J. (1972) *Nature* **237**, 29.
BRAES, L. L. and MILEY, G. K. (1972) *Ibid.* **235**, 273.
COHEN, M. (1973) *Mon. Not. Roy. astr. Soc.* **161**, 97.
CROMPTON, D. and HUTCHINGS, J. B. (1972) *Nature* **237**, 92.
EZER, D. and CAMERON, A. G. W. (1965) *Can. J. Phys.* **43**, 1487.
GAPOSCHKIN, S. (1937) *H.A.* **105**, 515.
HARO, G. and CHAVIRA, E. (1965) *Vistas in Astronomy* **8**, 89.
KEMP, J. C. (1969) *J. Opt. Soc. Amer.* **59**, 950.
KEMP, J. C. and WOLSTENCROFT, R. D. (1972) *Astrophys J. Lett.* **176**, L115.
KNACKE, R. F., STROM, K. M., STROM, S. E., YOUNG, E. and KUNKEL, W. (1973) *Astrophys. J.* **129**, 847.
KUKARKIN, B. V. and PARENAGO, P. P. (1948) *General Catalogue of Variable Stars*, First Ed., Moscow.
KUKARKIN, B. V., PARENAGO, P. P., EFREMOV, YU. I. and KHOLOPOV, P. N. (1958) *Ibid.* Second Ed., Moscow.
LARSON, R. B. (1972) *Mon. Not. Roy. astr. Soc.* **157**, 121.
ROSINO, L. (1969) *Non-Periodic Phenomena in Variable Stars*, p. 173, Reidel, Dordrecht.
SHAPLEY, H. (1924) *Harvard Bull.* No. 803.
STOKES, R. A., AVERY, R. W. and MICHALSKY, J. J. JNR. (1973) *Nature Phys. Sci.* **241**, 5.
STROM, K. M., STROM, S. E. and YOST, J. (1971) *Astrophys. J.* **165**, 479.
STROM, S. E., STROM, K. M., BROOKE, A. C., BREGMAN, J. and YOST, J. (1972a) *Ibid.* **171**, 267.
STROM, S. E., STROM, K. M., YOST, J., CARRASCO, L. and GRASDALEN, G. (1972b) *Ibid.* **173**, 353.
VAN DEN BERGH, S. (1972) *Nature* **235**, 273.

CHAPTER 11

Spatial distribution

LIKE the RW Aurigae variables discussed earlier, the stars of the T Orionis class appear to be somewhat more widespread in their spatial distribution than the T Tauri variables. For example, we find a large concentration of these stars in the Milky Way regions of Perseus, Auriga, Taurus, Orion and Monoceros, and a second wide area banded along the galactic equator from Ophiuchus, through Scorpius, Corona Australis and Carina.

Smaller numbers have also been discovered in the constellations of Cepheus and Vulpecula. The one common feature is that all lie close to the galactic equator where we would expect to find large regions of dark and bright nebulosity.

The majority of these variables which are known at present are located within a radius of a few hundreds of parsecs from the Sun. Table IX gives the approximate galactic coordinates of several of the larger associations of T Orionis variables.

In many cases, the T Orionis stars are found in close proximity to both RW Aurigae and T Tauri variables and, as has been pointed out earlier, it is often difficult to assign a faint nebular variable unambiguously to any particular class due to the inevitable paucity of both photometric and spectroscopic observations.

Where the variable is sufficiently bright, these difficulties are not so pronounced since detailed light curves are available for study.

TABLE IX. GALACTIC COORDINATES OF T ORIONIS ASSOCIATIONS

Constellation	Galactic coordinates*	
	l^{II}	b^{II}
Cepheus	75°	+10°
Perseus	118°	− 5°
Orion	174°	−18°
Orion	178°	−20°
Scorpius	318°	+15°
Ophiuchus	325°	+ 5°
Corona Australis	328°	−20°
Sagittarius	333°	− 2°
Sagittarius	339°	+ 1°

*Epoch 1950.0.

T Orionis Variables in Specific Regions

IV CEPHEI

The T-association IV Cephei has been described by Blanco and Williams (1959). This aggregate contains several faint Hα emission line objects discovered by Blanco (1962) which are embedded in a very irregular absorption cloud. A handful of variables which appear to be T Orionis stars have also been located in this region, all of which are extremely faint objects with photographic magnitudes between 13ᵐ and 17ᵐ.

PERSEUS REGION

Some T Orionis variables and also bright A and B type stars of the XX Ophiuchi group, have been found in heavily obscured regions of Perseus, particularly around δ Perser. Two of these early type stars (X and XY Persei) are among the brightest of all these variables (see Table VI).

THE TRAPEZIUM REGION IN ORION

Numerous T Orionis variables have been found in the vicinity of the Trapezium (θ Orionis) during surveys carried out by various workers including Rosino (1946, 1956, 1962, 1969) and Blanco (1963). This particular region contains the nebulae NGC 1976, 1977, 198_ and 1999 and also several very heavily obscured areas where the variables are to be found mostly within the fringes of dark nebulosities.

Near the Trapezium itself, the density of T Orionis variables is remarkably high with 70 per cent of all the stars in this region belonging to this, or the T Tauri, class. The total number of variables found to the end of 1972 in this area is 493.

A small number of variables examined by Brun (1934) and later by Parenago (1954) have M type spectra and, although classified as T Orionis variables by Kukarkin and Parenago (1948), are almost certainly T Tauri stars. These include such variables as SZ Orionis (spectrum M2, range 14.4–16.2 m_{pg}) and BS Orionis (spectrum M2, range 15.3–16.5 m_{pg}). Not only are their amplitudes small, more in keeping with members of the T Tauri class, but being faint objects, few details of their light variations are available.

Typical members of this class found near the Trapezium are XX Orionis (14.2–16.3 m_{pg}) and BZ Orionis (14.8–16.4 m_{pg}) both of which were discovered by Shapley (1924) during one of the early photographic surveys of this region.

THE HORSEHEAD NEBULA AND NGC 2024

This dark cloud lies close to the second magnitude star ζ Orionis and is probably closely associated with the foregoing region since Rosino (1962) has found a narrow lane of variables running from NGC 1977 and NGC 1999 towards the Horsehead Nebula.

OPHIUCHUS–SCORPIUS REGION

A number of T Orionis variables have been found in a complex area of the Milky Way where there are several regions of dark and bright nebulosity, particularly near NGC 6273 and NGC 6121. Typical members are XX and IX Ophiuchi, and AT and AU Scorpii (see Table VI).

XX Ophiuchi has sometimes been regarded as the prototype of a small subgroup of stars very similar to the T Orionis variables in their general light variations. They are usually white variables with A or B type spectra whose brightness fluctuations are characterized by non-periodic minima.

CORONA AUSTRALIS REGION

A group of bright T Orionis stars are situated in a small region in the vicinity of NGC

PLATE III. The Region around NGC 2023-24. The bright star is Orionis. (After Kerolyr, Forcal quier.)

6723, in the same field as γ Coronae Australis. Six of these were described by Gaposchkin (1937a, 1937b) many years ago and typical light curves for the four brightest members of the group are given in Figs. 6–9. The two remaining variables are VV Coronae Australis with an amplitude of more than 4^m (13.0– < 17.0 m_{pg}) and DG Coronae Australis (13.1–15.3 m_{pg}).

More recently, the stellar group around R Coronae Australis has been examined by Strom *et al.* (1972). The dark cloud in this region appears to have been the site of active star formation in the very recent past. Apart from the variables just mentioned, there are a significant number of Hα emission objects which have been identified from objective prism plates. Optical and infrared observations of 11 of these objects have shown that most of them are surrounded by gas and dust envelopes. While the distribution of stellar luminosities in this region appears to be very similar to that found for the Taurus cloud, it is noticeable that the optical depths of the circumstellar clouds are appreciably larger for the Corona Australis variables. From this it may be inferred that these stars are somewhat younger than their Taurus counterparts. The age of this aggregate has been estimated at $\sim 10^6$ years.

Knacke *et al.* (1973) have suggested that these variables have masses in the range 1–4 $M_\odot$, not unlike the mass range of similar stars in Taurus.

THE SAGITTARIUS RIFT

Although the Aquila rift shows a decided paucity of H emission stars and other nebular variables, a small number have been found in the similar region in Sagittarius with galactic coordinates $l^{II} = 325–340°$; $b^{II} = \pm 6°$.

Most of these objects are quite faint and further observational data concerning their light variations and spectroscopic characteristics are required before they can be definitely assigned to the T Orionis or other nebular variable class. DU Sagittarii (13.2–16.6 m_{pg}) has been classed as a T Orionis star by Himpel (1944).

NGC 7000 AND IC 5070 REGION

Apart from the relatively large groupings of T Orionis variables mentioned above, other small aggregates have also been located in various regions. Welin (1973) has carried out a search for Hα emission stars on objective prism plates taken with the Uppsala–Kvistaberg Schmidt telescope covering a region in Cygnus comprising NGC 7000 and IC 5070. A total of 141 stars in this category were discovered, of which 35 had been previously investigated. Welin has prepared a catalogue of these stars giving their positions, visual and blue magnitudes, Hα intensities and, where known, their spectral types.

MONOCEROS REGION

A region in Monoceros with galactic coordinates $l^{II} = 203°$, $b^{II} = \pm 2°$ has been examined by Karlsson (1972). The space densities of types O to F7 stars are given (some of which may be early type T Orionis variables), together with the general behaviour of the interstellar extinction. The open cluster NGC 2264 lies in the region covered. As we shall see in Chapter 17, this area is particularly rich in T Tauri variables.

R-ASSOCIATIONS

Racine (1970) has discussed the spectroscopic and photometric characteristics of certain BD stars in reflection nebulae. On the basis of the galactic coordinates and distances of the illuminating stars, the reflection nebulae are classified into 15 groups known as

The Nebular Variables

R-associations. The galactic distributions show close similarities with those of OB-associations lying within 1 kpc of the Sun.

The absolute magnitudes determined for 7 emission B type stars in R-associations are found to be $\sim 1^m$ fainter at a given $(V-B)_0$ than those of Be stars in the general field. Racine has suggested that the Be stars in reflection nebulae may be rotationally unstable objects still in the pre-main sequence stages of their evolution. A large number of peculiar A and B type stars found in reflection nebulae also have appreciable ultraviolet excesses and the occurrence of these objects in young stellar groups implies that some of the early type stars found in the R-associations may be massive counterparts of the T Tauri variables. Although few details are available concerning the variability of these stars, it is possible, taking into account their masses and spectra, that they may be T Orionis variables.

The close association of T Orionis variables and OB type stars as far as their galactic distributions are concerned has also been found by Artiukhina (1970). Two groups of T Orionis variables in Taurus and a group of OB stars of the Cepheus II association both show similar velocities towards the galactic plane. Similarly, the V_z values for several groups of T Orionis stars in Orion and stars in the ζ Persei association are close to zero.

References

ARTIUKHINA, N. M. (1970) *Astron. Zh.* **47**, 667.
BLANCO, V. M. and WILLIAMS, A. D. (1959) *Astrophys. J.* **130**, 482.
BLANCO, V. M. (1962) *Publ. astr. Soc. Pacif.* **74**, 330.
BLANCO, V. M. (1963) *Astrophys. J.* **137**, 513.
BRUN, A. (1934) *Publ. Obs. Lyon* **3**, 149.
GAPOSCHKIN, S. (1937a) *H.A.* **105**, 514.
GAPOSCHKIN, S. (1937b) *Ibid.* **105**, 515.
HIMPEL, K. (1944) *Beobach. Zirkular der Astron. Nachr.* **26**, 25.
KARLSSON, B. (1972) *Astron. & Astrophys.*, Suppl. Ser. **7**, 35.
KNACKE, R. F., STROM, K. M., STROM, S. E., YOUNG, E. and KUNKEL, W. (1973) *Astrophys. J.* **179**, 847.
KUKARKIN, B. V. and PARENAGO, P. P. (1948) *General Catalogue of Variable Stars*, Vol. 1, Moscow.
PARENAGO, P. P. (1954) *Trudy. Sternberg Astr. Inst.* Vol. 25.
RACINE, J. A. R. (1970) *Thesis*, Univ. of Toronto, Ontario.
ROSINO, L. (1946) *Publ. Bologna* V, No. 1.
ROSINO, L. (1956) *Mem. Soc. astr. Ital.* **27**, 3.
ROSINO, L. (1962) *Ibid.* **32**, 4.
ROSINO, L. (1969) *Non-Periodic Phenomena in Variable Stars*, p. 173, Reidel, Dordrecht.
SHAPLEY, H. (1924) *Harvard Bull.* 803.
STROM, K. M., STROM, S. E., KNACKE, R. F. and YOUNG, E. (1972) *Bull. Am. astr. Soc.* **4**, 325.
WELIN, G. (1973) *Astron. and Astrophys. Suppl. Ser.* **9**, 183.

CHAPTER 12

Interaction with nebulosity

THE INTIMATE physical relationship which exists between many of these variables and their parent nebulae is clearly of great importance in the problem of understanding the evolutionary processes going on both in the stars themselves and in the surrounding dust clouds. Some of the nebulae associated with these variables are known to vary in brightness and the exact relation between the variability of star and nebula has still to be fully resolved.

From the evolutionary characteristics of the nebular variables as a whole it seems clear that the main interaction of the nebula with the stellar core will be in the form of infalling material which may still be present once the star reaches the main sequence. The amount of such material at this stage will, of course, depend to a large extent upon the initial mass of the protostellar cloud and that of the star.

Local variations in the density of solid particles which form circumstellar shells around these stars will also have an effect upon the light fluctuations of these variables. Since these changes occur on a relatively short time scale we may obtain important data as to the evolution of these shells from a close study of the different types of irregular light variations found in these stars.

In certain cases, too, particularly where the stellar core has a high temperature and luminosity (as we find in several T Orionis variables) another effect comes into play. Here, radiation pressure will exert a definite force upon the surrounding cloud. In extreme cases, this may even be sufficiently strong to disrupt the cloud permanently, giving rise to an object such as FU Orionis or V1057 Cygni.

Interaction with Infalling Material

The high kinetic energy of the infalling material will produce a high temperature just inside the shock front surrounding the stellar core as described earlier in Chapter 6 for the RW Aurigae variables. The treatment of the shock front given there is also generally applicable in this case, Larson (1969). As before, therefore, we normally find that these variables possess a higher luminosity than main sequence stars having the same spectral type.

In its early evolutionary stages of collapse, the stellar core will be completely obscured by the circumstellar shell and, since the greater proportion of the radiation will be infrared radiation produced by thermal emission from the dust grains, this will obliterate any evidence of infalling material. During its final approach to the main sequence, however, once the stellar core has accreted sufficient material from the surrounding cloud to enable visible radiation to emerge, the effect of infalling material on the brightness of the star, and in its spectrum, might be expected to show itself.

The Nebular Variables

Much will depend, of course, upon the time scale of this particular phase. The transition stage between emergence of visible radiation and accretion of most of the surrounding material will be relatively short. We would, therefore, expect to find observational evidence of infalling material only in a comparatively small number of the youngest variables.

Such evidence for a variable infall of material onto the surface of certain T Orionis (and T Tauri) stars has been provided by Walker (1961, 1963, 1964, 1966 and 1969). A number of T Orionis variables in the Orion Nebula display an ultraviolet excess but, unlike the majority of such variables, these also have an inverse P Cygni absorption spectrum. The prototype star of this subgroup is YY Orionis ($13^m.9$–$15^m.7$ photographic).

In the spectra of these stars we find emission lines which possess a radial velocity very similar to that of the aggregate to which they belong, whereas the absorption components are shifted to longer wavelengths with radial velocities up to 400 km/sec more positive than the cluster velocity. These absorption lines are always those of hydrogen and sometimes of Ca II.

The presence of this material falling onto the stellar surface would be expected to cause a brightening of the star and, if the amount varies with time, can explain some of the irregular light variations. Since it is also found to occur only in those variables which have a large ultraviolet excess, it would also appear to be intimately connected with this effect.

This correlation between visual magnitude and the amount of infalling material is confirmed by the data given in Table X for XX Orionis; similar data for SU Orionis is given in Table XIII (Chapter 15).

Although only three plates are available for XX Orionis, it appears that the "YY Orionis" effect does, indeed, disappear when the variable is around minimum light. Taken in conjunction with the data in Table XIII, there seems little doubt that, at least in SU and XX Orionis, the amount of infalling material in these systems is variable and directly linked with the visual brightness.

TABLE X. SPECTROSCOPIC AND PHOTOMETRIC OBSERVATIONS OF XX ORI

Plate No.	Date (UT)	Exp (min)	Radial velocity (km/sec)		Magnitude (visual)
			Emission	Absorption	
ECL-218*	21 Nov. 1962	120	$+50\pm5$	—†	14.6
ES-312	22 Nov. 1962	124	$+20\pm6$	$+333\pm9$	14.6
ES-402	27 Feb. 1963	127	-6 ± 3	—‡	15.1

*Plate taken with the Lallemand electronic camera and coudé spectrograph; dispersion 48 Å/mm.
†Absorption spectrum present but too weak to measure.
‡Absorption spectrum definitely absent.

RADIATION PRESSURE EFFECTS

Most of the T Orionis variables investigated by Walker show the normal P Cygni type of spectrum indicating that matter is moving radially outward from the star into the surrounding medium. In terms of evolutionary progress we may consider such variables to have accreted most of the protostellar cloud in the vicinity of the star and, possibly, to be slightly older than those just discussed.

In many stars, particularly those with early spectral types, radiation pressure will exert a definite effect upon the surrounding gas and dust grains. Gilman (1972) has shown that while these grains are not position-coupled to the gas, they are momentum-coupled. The radiation pressure therefore not only drives the mass loss from these variables, but will have the effect of dispersing the obscuring material. The variable absorption effect, produced by changes in the density of such particles, can be taken for the interpretation of some of the light variations.

INTERSTELLAR EXTINCTION IN THE ORION ASSOCIATION

The fact that the hot, young stars of the T Orionis type modify the obscuring matter in their neighbourhood has been demonstrated by Lee (1968). Photoelectric photometry of 196 members of the I Orionis association was obtained in order to ascertain the wavelength dependence of the interstellar extinction throughout the entire association.

The regional extinction laws were calculated from the colour excesses of individual stars in certain localized areas while equivalent monochromatic laws were also computed. It was discovered that two regions of the I Orionis association exhibit significantly "non-normal" extinction. These are in the centre of the Orion Nebula and in the east "belt" region near ζ Orionis.

In both cases, the anomalies are a high ratio of total to selective absorption and a very steep infrared extinction gradient. In the region around ζ Orionis, the ratio of total to selective absorption, $R \simeq 5.5$, whereas near θ^1 Orionis, $R > 5.5$. In the Orion Nebula, this anomaly appears to be confined to the centre of the nebula within $15'$ of θ^1 Orionis.

All attempts to explain the very strong infrared radiation of the early type stars in this region in terms other than interstellar extinction have been unsuccessful. Circumstellar emission, while certainly quite possible, cannot account for the entire effect. From his observations, Lee has found that in all cases, the stars in I Orionis which exhibit decidedly non-normal extinction laws are intimately associated with bright nebulosity and are of spectral types B and A. Curiously, similar type stars in the same region often exhibit no such anomalies.

This data is interpreted as providing additional confirmation of the idea that these high-temperature, pre-main sequence stars modify the surrounding gas and dust clouds to a considerable extent. Several of the extinction laws found for stars in this association closely resemble those which result from bimodal distributions of particles having different characteristic dimensions.

Variable Nebulae associated with T Orionis Variables

The number of variable nebulae known at the present time is very small compared with the total number of nebular variables. In spite of the fact that many of these stars are situated in, or close to, small reflection nebulosities, the majority of the latter appear to be constant in brightness.

This does not mean to say that variations in the brightness of the exciting star have no effect at all on the surface luminosities of these nebulae. It is much more likely that any changes in brightness have passed unnoticed for two reasons.

First, it is only in the case of the brightest of these objects that a sufficient number of observations, visual and photographic, have been made with the intention of detecting any

The Nebular Variables

variability. Second, the amplitudes of many of these stars are often quite small ($\Delta m_v < 1^m.0$) and the corresponding changes in the light of the nebulae may be at the limit of detection.

In the case of S Coronae Australis, however, which illuminates the nebulosity NGC 6729, very definite variations in brightness have been recorded for this nebula. These have been summarized by Knox-Shaw (1924). As yet, it is not possible to state whether these variations in the brightness of NGC 6729 are related to those of S Coronae Australis or if the two are completely independent of each other.

References

GILMAN, R. C. (1972) *Astrophys. J.* **178**, 473.
KNOX-SHAW, H. (1924) *Helwan Bull.* **2**, 76.
LARSON, R. B. (1969) *Mon. Not. Roy. astr. Soc.* **145**, 289.
LEE, T. A. (1968) *Astrophys. J.* **152**, 913.
WALKER, M. F. (1961) *Comptes Rendus* **253**, 383.
WALKER, M. F. (1963) *Astrophys. J.* **68**, 298.
WALKER, M. F. (1964) *Roy. Obs. Bull.* No. 82, 69.
WALKER, M. F. (1966) *Stellar Evolution*, p. 405, Plenum Press, New York.
WALKER, M. F. (1969) *Non-Periodic Phenomena in Variable Stars*, p. 103, Reidel, Dordrecht.

CHAPTER 13

Evolutionary characteristics

HAVING discussed a possible evolutionary scheme for protostars with masses of $\sim 5\,M_\odot$ earlier, in Chapter 7, we shall here examine the non-homologous and homologous collapse of a protostellar cloud to give a protostar of approximately 2.5 $M_\odot$. As was pointed out earlier, this differentiation of the three classes of nebular variables according to their masses cannot be precisely defined and the available evidence for this is, to say the least, very slender.

Nevertheless, as Larson (1972a) has suggested, R Coronae Australis, which we include among the T Orionis variables on the basis of its light variations, can be interpreted as a protostar with a total mass of approximately 2.5 $M_\odot$, virtually all of which is concentrated in the central star. This variable, therefore, is very similar in many respects to R Monocerotis although it is somewhat fainter and its infrared spectrum peaks at a shorter wavelength. The position of these two stars on the infrared Hertzsprung–Russell diagram may be compared in Fig. 19.

Here, we must not lose sight of the fact that, like the RW Aurigae variables, the T Orionis stars cover a wide range of spectral types and it seems reasonable to suppose that their luminosities also vary over a similar range. We shall also assume that their masses lie between 2 and 3 $M_\odot$ as extreme values. How valid this assumption is for all T Orionis variables is something which must await further observational data.

Non-homologous Collapse

GENERAL ASSUMPTIONS

We assume, following Larson (1969, 1972a), that the initial and boundary conditions of the protostellar cloud are the same as those previously considered in Chapter 7. Thus the initial temperature will be taken as 10°K, the hydrogen is assumed to be all in the molecular form and a constant infrared opacity due to the presence of dust grains of 0.15 cm² g⁻¹ is taken.

Gaustad (1963) has considered a similar situation and shown that the dust grains constitute ~ 1 per cent of the total material in the protostellar cloud. This is compatible with star formation occurring in an H I region having a chemical composition which is typical of Population I objects.

The Rosseland mean opacity: $\kappa_R = 0.15$ cm² g⁻¹ given above would certainly appear to hold at temperatures below the evaporation temperature of the dust grains. Above this temperature (~ 1400°K), molecular absorption becomes a dominant factor and the opacity falls by an appreciable amount. Cox (1966) has calculated opacities for T $\leqslant$ 1500°K, while the effect of molecular absorption on the opacity for temperatures up to 3000°K has been elucidated by Tsuji (1966).

The Nebular Variables

The evolution of the stellar core for the cases where the mass is 2.0 and 3.0 $M_\odot$ has been computed by Larson (1969, 1972a) and the relevant curves are shown in Figs. 31 and 32. As before, the dashed portion along the curve represents the extremely short initial period when all of the radiation from the core is absorbed by the infalling material. The dot indicates the point where half of the mass has been accreted by the core and the open circle, the point where either all of the mass has been accreted and the visual optical depth of the cloud is approximately unity (Fig. 31) or where the stellar core is essentially on the main sequence (Fig. 32).

Comparison with the similar diagrams for the RW Aurigae and T Tauri variables shows that the case where the mass is 2.0 M represents a transition point between low and high mass protostars.

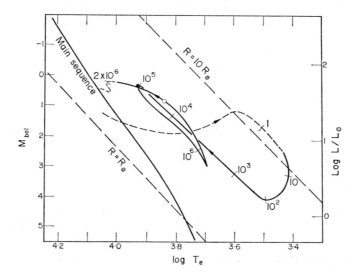

FIG. 31. The evolution of the stellar core for a protostar of 2 $M_\odot$. The time in years following the formation of the core is marked along the curve. The solid dot represents the point where half of the mass has been accreted and the open circle where essentially all of the mass has been accreted and the core becomes visible as a pre-main sequence object. (After Larson.)

Throughout most of the accretion process, the evolution of a 2 $M_\odot$ protostar follows a very similar course to that of the less massive protostars. A change occurs, however, after a period of about 1.2×10^6 years following the formation of the core. By this time, almost all of the mass has been accreted by the central core. The infalling material, though, still possesses a mass of approximately 0.006 $M_\odot$ and has an optical depth of ~ 3.

This means that radiative energy transfer now becomes dominant in the inner regions of the core which have a temperature of close on $9 \times 10^{6\circ}$K. Now this process transports an increasingly significant amount of energy through the core to the outer levels. We thus have a luminosity wave travelling outward through the core, causing the outer regions to expand quite rapidly. Indeed, this expansion appears to take place on a time scale which is comparable with the thermal relaxation time, being of the order of 1 day near the surface. Owing to the accompanying rapid brightening of the core, the luminosity doubles in a period of about 4 days from 15 to 26 $L_\odot$.

Following this phase, the star then settles down to a more leisurely evolutionary process

84

since it is now in radiative equilibrium and it continues to evolve towards the main sequence along a radiative track. While the luminosity during this phase increases, the radius decreases. After about 1.4×10^6 years following the formation of the core, we find that the optical depth of the remainder of the protostellar cloud has dropped below unity.

In the case of the 3 $M_\odot$ protostar, the course of evolution follows a similar pattern with a luminosity wave travelling radially outward through the core, bringing about the accompanying rapid expansion of the outer regions.

The major difference here is that there is now relatively more material still present in the protostellar cloud and, since most of the luminosity still comes from the kinetic energy of this material as it reaches the shock front, the luminosity falls rather than increases. This is shown by the dip in the evolution curve in Fig. 32 at about 10^5 years.

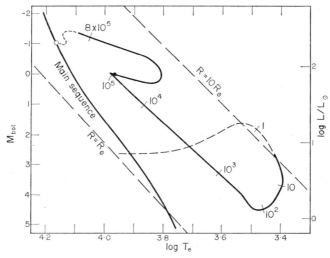

FIG. 32. The evolution of the stellar core for a protostar of 3 $M_\odot$. Here the open circle represents the point where the stellar core first evolves into the main sequence. (After Larson.)

Once the core has settled into radiative equilibrium, contraction towards the main sequence is similar to the previous case although here there is still accretion of mass from the remnants of the protostellar cloud. The stellar core, therefore, continues to gain mass during this pre-main sequence phase. The optical depth of the remaining material in the proto-stellar cloud for a 3 $M_\odot$ protostar by the time it reaches the main sequence has been calculated by Larson (1972a) to be ~ 3. This figure may be higher than the actual value, particularly if the dust grains are evaporated or removed by other effects at temperatures $\sim 2000°$K.

Since the mass of the remaining material around the central core for a star of ~ 3 $M_\odot$ is of the order of 0.04 $M_\odot$, this may represent a limiting value in the sense that more massive protostars will contain sufficient infalling material by the time they reach the main sequence for this to heavily obscure the star. Consequently, it is unlikely that many pre-main sequence stars much more massive than 3 $M_\odot$ will be optically visible. Only such objects as R Monocerotis which probably have masses of the order of 5 $M_\odot$ have been observed and here, as we have seen, much of the radiation of the central star is drastically modified by the surrounding dust shell.

The Nebular Variables

Comparison with Observation

Assuming the above evolutionary scheme to be essentially correct, and on the basis of a mass range from ∼2 to 3 $M_\odot$, we would expect to find a variety of effects among the T Orionis variables, depending upon the mass of the infalling cloud and its optical depth during the pre-main sequence stage when the star becomes optically visible.

From the light curves given in Chapters 2 and 8, we see that there are non-periodic minima, very often with amplitudes of $2^m.0$ to $2^m.5$, together with smaller, but still quite pronounced, fluctuations around maximum brightness. Both types of light variation occur on a time scale of 5–10 days and seem to be related to two characteristics of these stars.

(a) Variations in the amount of infalling material and its optical depth. The latter, in turn, depends upon the depletion of the dust grains either by evaporation or other effects. In the case of those stars with early type spectra we would expect temperatures sufficiently high to evaporate a greater proportion of the dust grains (graphite grains, for example, have an evaporation temperature of ∼2000°K) compared with those T Orionis variables with later type spectra. Radiation pressure, too, will play a role in dispersing some of the surrounding material, particularly in the XX Ophiuchi variables of spectral types Be and A. This will lead to more of the radiation from the stellar core shining through the enshrouding dust.

(b) There appears to be little doubt that very violent convective and chromospheric activity takes place in the atmospheres of these stars during the pre-main sequence stage of their evolution. This could result in the appearance of the rapid fluctuations in brightness around "normal" light, typical of such variables as SY, XX, YY, YZ and HS Orionis. These fluctuations occur on a time scale of the order of minutes or a few hours. Such "flare-like" activity must not, however, be confused with the flares observed in the Orion stars such as those previously discussed in Chapter 8.

In the case of the increases in brightness around maximum light (time scale = 5–10 days), it is perhaps significant that these are more prominent in those stars possessing early type spectra. If we assume that these variables are relatively more luminous than the stars with later type spectra, radiation pressure will start to become more dominant over gravity once the infalling cloud becomes optically thin.

Larson and Starrfield (1971) have shown that in an optically thin medium, radiation pressure is dominant if

$$\frac{L/L_\odot}{M/M_\odot} \geqslant 50 \tag{1}$$

which is true for masses in excess of 3 $M_\odot$.

In the case of those objects where the central star ejects material (which is true of most of the T Orionis variables) this will also add to the disruption of the infalling cloud and, when the mass of the central star is well in excess of 3 $M_\odot$, this could lead to the total, and permanent, removal of the surrounding cloud. The result would be a rapid brightening of more than 5^m as has been found for such objects as FU Orionis and V1057 Cygni.

Homologous Collapse

As in the case of the RW Aurigae variables, the use of homologous models to describe the pre-main sequence evolution of the stars we are considering here will lead to inaccuracies since the mass considered is too large for them to remain fully convective during their pre-main sequence lifetimes. A criticism of the use of homologous models has been made by Iben (1965) although he has pointed out that accurate results can be obtained by this method so long as the models are fully convective.

Only for protostars with masses $< 1.5\,M_\odot$ does the resulting star possess any sizeable outer convection zone. Such a star first appears very near the bottom of its "Hayashi" track as shown by Hayashi, Hoshi and Sugimoto (1962). We would therefore expect stars with masses in the range we are discussing here to have no "Hayashi" track since the entropy and radius are never sufficiently large for a normal Hayashi phase to have any existence.

Effect of Rotation

A variety of methods have been used in recent years to investigate the static structure of rotating stars, for example those of Faulkner et al. (1968), Sackman and Anand (1970) and Kippenhahn and Thomas (1970), while the problems of rotating stars have been examined by Kippenhahn et al. (1970) and Moss (1973).

In addition, several authors have calculated the early stages of a non-rotating protostellar cloud, for example Bodenheimer and Sweigart (1968), Disney et al. (1969), Hunter (1969), Larson (1969) and Penston (1969), while Larson (1972b) has examined the collapse of a rotating cloud.

Most of these investigations have been carried out for protostars of $\leqslant 1\,M_\odot$, however, and are discussed in detail later, in Chapter 19. Here, we shall confine ourselves to an examination of the effect of rotation and mass loss upon the contraction times of these variables and, in particular, the rapidly rotating Be type variables of the XX Ophiuchi subgroup.

In the case of these variables particularly, it appears that one of the most efficient mechanisms for losing angular momentum during contraction is that of equatorial mass shedding. Following Moss (1973) we have

$$L = 0.48\,\frac{JGM^2}{R_e} \tag{2}$$

where L is the total energy flux of the star, J is the contraction parameter, G is the gravitational constant, M is the mass and R_e is the equatorial radius.

Calculations show that, in general, rapidly rotating models take a longer time to contract towards the main sequence than non-rotating ones. Where equatorial mass shedding is allowed, the contraction time is reduced by approximately 20 per cent but the contraction time scale is still appreciably longer than for a non-rotating star of the same mass. It must be appreciated, however, that these calculations have not, as yet, been carried through fully for stars with masses much greater than $1\,M_\odot$ and caution must be used when applying them to these particular variables although it seems quite possible to apply them to the T Tauri stars with smaller masses.

References

BODENHEIMER, P. and SWEIGART, A. (1968) *Astrophys. J.* **152**, 515.
Cox, A. N. (1966) Unpublished tables of opacities supplied privately to P. R. Demarque.
DISNEY, M. J., McNALLY, D. and WRIGHT, A. E. (1969) *Mon. Not. Roy. astr. Soc.* **146**, 123.

87

The Nebular Variables

FAULKNER, J., ROXBURGH, I. W. and STRITTMATTER, P. A. (1968) *Astrophys. J.* **151**, 203.
GAUSTAD, J. E. (1963) *Ibid.* **138**, 1050.
HAYASHI, C., HOSHI, R. and SUGIMOTO, D. (1962) *Prog. Theor. Phys. Suppl.* No. 22.
HUNTER, C. (1969) *Mon. Not. Roy. astr. Soc.* **142**, 473.
IBEN, I. (1965) *Astrophys. J.* **141**, 993.
KIPPENHAHN, R. and THOMAS, H. C. (1970) In *Stellar Rotation*, SLATTERBAK, A. ed., Reidel, Dordrecht.
KIPPENHAHN, R., MEYER-HOFMEISTER, E. and THOMAS, H. C. (1970) *Ibid.*
LARSON, R. B. (1969) *Mon. Not. Roy. astr. Soc.* **145**, 271.
LARSON, R. B. and STARRFIELD, S. (1971) *Astron. and Astrophys.* **13**, 190.
LARSON, R. B. (1972a) *Mon. Not. Roy. astr. Soc.* **157**, 121.
LARSON, R. B. (1972b) *Ibid.* **156**, 437.
MOSS, D. L. (1973) *Ibid.* **161**, 225.
PENSTON, M. V. (1969) *Ibid.* **145**, 457.
SACKMAN, I. J. and ANAND, S. P. S. (1970) *Astrophys. J.* **162**, 105.
TSUJI, T. (1966) *Publ. astr. Soc. Japan* **18**, 127.

PART III

T Tauri Variables

CHAPTER 14

Light variations in T Tauri stars

THE CLASSIFICATION of the T Tauri variables as a distinct group of the nebular variables was originally made by Joy (1945) who described eleven stars which had, as their major characteristics, the Hα line in emission and close association with nebulosity. Herbig (1960) has defined the following characteristics of these stars; spectral type G–K, low to moderate luminosity, variable emission line spectrum and intimate connection with a dark nebulae. While most of these definitions appear satisfactory, we may see from the data in Table XI, that the majority of the fainter T Tauri stars have spectra of Type M.

T Tauri ($9^m.5$–$13^m.0$; spectrum dG5e) is one of the brightest and most representative of this group.

Several objective prism surveys have been carried out since 1945, in a search for short spectra showing the Hα line in emission, resulting in the discovery of 126 such variables by 1962. The total number is now well over 200 (1972).

The importance of these stars lies mainly in the fact that they all appear to be very young objects still in the process of gravitational contraction from the diffuse gas and dust clouds in which they are embedded. Since several of these variables are extremely faint, even at maximum brightness, it is likely that some of them will be found to belong to the associated RW Aurigae and T Orionis classes.

Light Curves of T Tauri stars

As with all of these variables, the light curves of the T Tauri variables show pronounced irregularities. In general, however, the light variations are not as rapid as those of the RW Aurigae or T Orionis variables, having the character of long, irregular waves. For many of these stars there are intervals of several hundreds of days when little sensible change is observed either visually or photographically.

T TAURI

In recent years, the prototype star has varied by only $0^m.5$, between $10^m.1$ and $10^m.6$ in the visible (Fig. 33). From a detailed examination of the light curve during this period we find that there are at least two components; a slow wave with a cycle length of approximately 100 days and an amplitude of $0^m.3$, upon which is superimposed a somewhat more rapid component with variations in brightness of $0^m.15$ and a cycle of between 1 and 5 days.

Some 20 years ago, the visual amplitude was much greater than it is now (Fig. 34) with the variable having an amplitude of $3^m.7$ making its extreme range $9^m.4$–$13^m.1$. During this period, the existence of a long cycle was not so pronounced as it is at present.

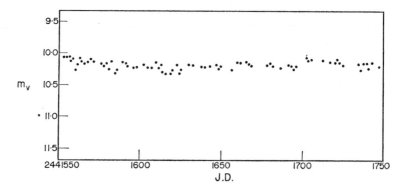

FIG. 33. Light curve of T Tauri showing the small amplitude of recent years.

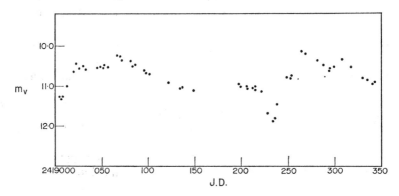

FIG. 34. Light curve of T Tauri showing the larger amplitude observed several decades ago.

Wenzel (1969) has carried out photoelectric measurements of this star on the UBV system which show that whereas the variations in the colour index (B–V) are only of the order of $0^m.05$, those of (U–B) are slightly greater (Fig. 36). In both the V/B–V and U–B/B–V diagrams, the continuous line represents the main sequence direction in this particular interval. The interstellar extinction is denoted by the arrow in the latter figure.

Comparison of the V/B–V and U–B/B–V diagrams for this star may be made with those of a typical T Orionis variable given earlier in Figs. 23 and 24. Unfortunately, no accurate series of three-colour observations are available for T Tauri during the period when its visual amplitude was in excess of three magnitudes.

RY TAURI

This is, perhaps, the brightest T Tauri variable known. The visual magnitude range given by Kholopov (1954) is $8^m.6$–$11^m.0$ and a photographic range of $10^m.1$–$12^m.3$. These values have been drastically revised since the very deep visual minimum of this star which occurred during 1959–60 (Fig. 37).

The V/B–V diagram for this star obtained by Wenzel (1969) shows certain peculiarities. The star moves vertically with (B–V) remaining almost constant and, when the variable is near maximum brightness, the motion is perpendicular to the direction of the main sequence.

TABLE XI. T TAURI VARIABLES

Star	Magnitude Maximum	Minimum	Spectrum
GM Aur	13.1	13.9	dK5e
GW Ori	10.8	11.4	dK3e
GX Ori	14.1	15.2	dK3e
HI Ori	13.1	15.5	dK0e
T Tau c	9.6*	13.5	dG5e
RY Tau c	8.6*	11.0	dG0e
UX Tau	10.7	13.4	dG5e
UZ Tau	11.7	14.9	dG5e
AA Tau	13.1	16.1	dM0e
BP Tau	12.3	13.3	dK5e
CW Tau	13.6	15.9	dK5e
CX Tau	14.5	15.3	dM1.5e
CY Tau	13.4	15.0	dM2e
CZ Tau	15.8	17.3	dM2e
DD Tau	14.5	15.5	dK6e
DE Tau	13.8	14.8	dM1e
DF Tau	11.9	13.8	dM0e
DH Tau	14.1	15.2	dM0e
DI Tau	14.0	15.0	dM0.5e
DK Tau	12.4	14.9	dM0e
DM Tau	14.5	15.2	dK5e
DN Tau	13.2	14.0	dK6e
DP Tau	13.7	15.4	dM0e
DQ Tau	14.0	15.0	dM0e
DS Tau	12.1	13.6	dK4e

*Visual magnitudes, all others being photographic.
c Associated with cometary nebulae.

CZ AND DD TAURI

This pair of T Tauri variables almost certainly form a binary system, the separation being 31″. The light curves of these stars (Fig. 39) are, again, quite typical of this class. Periods of virtually constant light are interspersed with long and irregular waves. CZ Tauri, in particular, is extremely faint. Throughout the entire period of observation, between October 1972 and March 1973, the visual range was between 14^m.8 and 15^m.6.

Owing to their general faintness and comparatively recent discovery, extensive and accurate light curves are available for only a few of the variables listed in Table XI. From a study of the existing data, it appears that there is no strict correlation between the visual or photometric determinations and spectral type. The fast and slow variations found for most of these stars seem to be completely irregular. Only very rarely do we find any evidence of cyclic changes in their light curves.

Light Variables of the Associated Nebulae

Several T Tauri variables are found associated with cometary nebulae; a small group of curiously-shaped diffuse nebulae, all of which appear to be in physical association with a

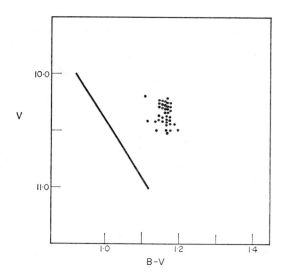

FIG. 35. V/B–V diagram for T Tauri. (After Wenzel.)

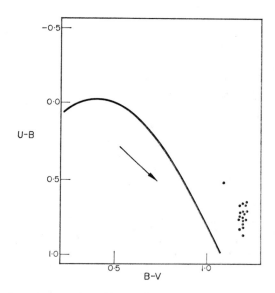

FIG. 36. Two-colour diagram for T Tauri. (After Wenzel.)

star. Often, but not always, these stars are variable and, as shown by Ambartsumian (1955), some of these belong to the T Tauri class.

Ambartsumian (1957) has also provided evidence that, like their allied stars, these diffuse nebulae are also very young formations. Four different types of cometary nebulae have been recognized according to their morphology. These are the biconical, conical and arcuate nebulae and those which are shown on long exposure photographs to be comma-like appendages to stars.

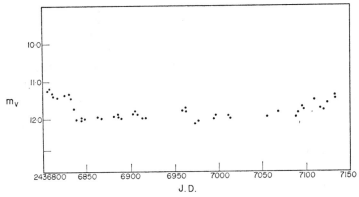

FIG. 37. Light curve of RY Tauri during the deep visual minimum of 1959–60.

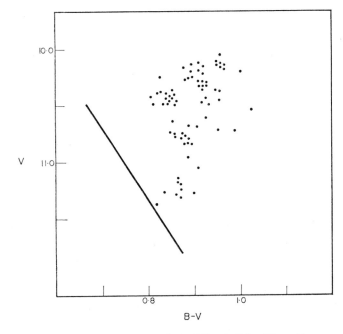

FIG. 38. V/B–V diagram for RY Tauri. (After Wenzel.)

The structure and brightness of some of the cometary nebulae are known to have varied, at times by quite appreciable amounts. Two of the brightest, NGC 1555 and Barnard B214 —associated with T and RY Tauri respectively—have been closely studied for several decades.

The small and complex emission-line nebula in which T Tauri is centrally situated may possibly be a Herbig–Haro object similar to those discussed in Chapter 22. Early visual observations, extending back to its discovery in 1852, indicate very significant variations in brightness. These, and more recent, determinations of the variability of this nebulosity have been summarized by Herbig (1950) and it is clear that the changes in brightness of the star and the nebula are not obviously related.

The Nebular Variables

The nebulosity surrounding RY Tauri consists of two components; a fan-shaped region lying to the north and west, and a much fainter region to the east of the star. Two 30-min exposures taken by Herbig (1961) with the Crossley reflector on 1 January 1957 and 18 March 1960 show considerable differences in the brightness of this nebula. The fan of nebulosity to the north and west had faded significantly in this period although the fainter region to the east showed a much smaller change. Of equal importance is the fact that RY Tauri itself passed through an abnormally deep minimum over the same period.

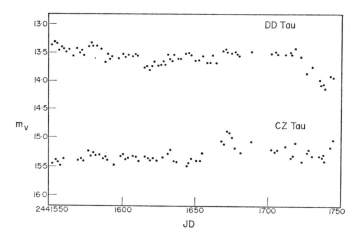

FIG. 39. Light curves of CZ and DD Tauri.

Infrared Observations

Recent theoretical studies of early stellar evolution, particularly those specifically directed towards an understanding of the pre-main sequence stage, have indicated that when such protostars are still surrounded by extensive gas and dust envelopes, phases of high luminosity are likely to occur. The fact that the T Tauri variables are stars still in this phase of their evolution was put forward by Herbig (1962).

While these stars are intrinsically faint in the visible region, the near infrared observations made by Mendoza (1966, 1968) indicated that the greater proportion of their radiation comes through in the infrared region beyond 2μ.

In addition to the T Orionis variables described earlier, Cohen (1973) has made multi-filter infrared observations of a series of T Tauri variables in the 2–22μ region. These observations were made with the Mount Lemmon 60-in. telescope using various photometers and doped germanium bolometers cooled to $2°K$. The internal sources of noise in the germanium bolometer were allowed for as described by Low (1961) and compared with sky (photon) noise. Table XII illustrates the multifilter data obtained by Cohen for two of these T Tauri variables on single nights, the values in parentheses being the 1 standard deviations of internal photometric errors.

From the magnitudes thus obtained, it is possible to define an energy distribution for both of the above stars (Fig. 40). The plot is one of wavelength (μ) against λF_λ in Watts cm^{-2} with both axes plotted logarithmically. In order to avoid overlapping, the curves are shifted in ordinate with one decade in λF_λ being shown.

96

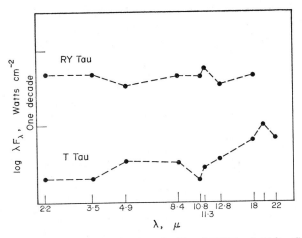

FIG. 40. Infrared energy distributions for T and RY Tauri. (After Cohen.)

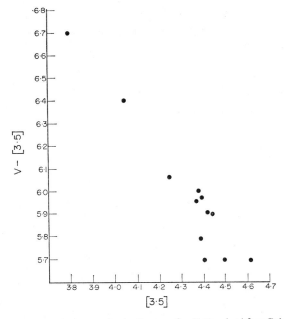

FIG. 41. Two-colour magnitude diagram for T Tauri. (After Cohen.)

It is immediately apparent that the energy distributions are both extremely flat and show no resemblance to the distributions which would arise from either free-free emission or the radiation from a black body at a single temperature. The most likely origin of this infrared radiation, based mainly upon the work carried out by Kuhi (1964) on the mechanism of mass loss from these young stars, is some form of solid circumstellar material. Whether this material is in the form of a shell or disc is still problematical.

From the light curves of the T Tauri variables we know that rapid fluctuations of the order of one magnitude or more may sometimes occur over periods of a few days and there may even be longer term variations. It is clearly of interest, therefore, to determine whether any

The Nebular Variables

TABLE XII. MULTIFILTER OBSERVATIONS OF T TAURI STARS

Narrow-band filter wavelength (μ)	T Tau	RY Tau
2.2	5.9 (0.1)	5.6 (0.1)
3.65	4.3 (0.1)	4.1 (0.1)
4.8	3.0 (0.2)	
8.6	1.1 (0.1)	1.5 (0.2)
10.8	0.9 (0.2)	0.8 (0.2)
11.3	0.4 (0.1)	0.5 (0.1)
12.8	−0.3 (0.1)	0.6 (0.1)
18	−2.0 (0.1)	−0.85 (0.3)
20	−2.6 (0.2)	
22	−2.5 (0.1)	

similar variations take place in the infrared. Mendoza (1968) has reported that certain of these stars show irregular fluctuations in brightness in all filters between 0.5 and 5.0μ. The evidence indicates that these are apparently in a similar time scale as the visible fluctuations.

Observations made during 1971–72 by Cohen (1973) on T and RY Tauri show significant changes at four effective wavelengths, namely 3.5, 4.9, 8.4 and 11μ. The two-colour-magnitude diagrams shown in Figs. 41 and 42 show both long- and short-term variations in the infrared for these variables. The long-term changes for T Tauri in Fig. 41 show a very definite trend with the near infrared flux increasing systematically as the star fades visually.

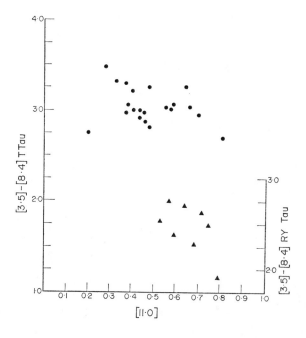

FIG. 42. Two-colour magnitude diagram for T and RY Tauri. (After Cohen.)

Herbig (1962) found a similar effect for RW Aurigae where the colour temperature decreased very steeply with a decrease in the visual brightness. Quite possibly the two effects are similar in nature

Short-term variations in the [3.5–8.4], [11.0] colour-magnitude diagram are clearly apparent for both T and RY Tauri. The trend is alike in both cases although the number of observations made of RY Tauri were smaller than for the former variable.

References

AMBARTSUMIAN, V. A. (1955) *Questions on Cosmogony* **4**, 76, Moscow.
AMBARTSUMIAN, V. A. (1957) *IAU Symposium No.* 3, *Non-Stable Stars*, Cambridge Univ. Press.
COHEN, M. (1973) *Mon. Not. Roy. astr. Soc.* **161**, 97.
HERBIG, G. H. (1950) *Astrophys. J.* **111**, 11.
HERBIG, G. H. (1960) *Astrophys. J. Suppl.* **4**, 337.
HERBIG, G. H. (1961) *Astrophys. J.* **133**, 337.
HERBIG, G. H. (1962) *Adv. Astr. Astrophys.* **1**, 47.
JOY, A. H. (1945) *Astrophys. J.* **102**, 168.
KHOLOPOV, P. N. (1954) *Variable Stars* **10**, 180.
KUHI, L. V. (1964) *Astrophys. J.* **140**, 1409.
LOW, F. J. (1961) *J. Opt. Soc. America* **51**, 1300.
MENDOZA, E. E. (1966) *Astrophys. J.* **143**, 1010.
MENDOZA, E. E. (1968) *Ibid.* **151**, 977.
WENZEL, W. (1969) *Non-Periodic Phenomena in Variable Stars*, p. 61, Reidel, Dordrecht.

CHAPTER 15

Spectroscopic characteristics

UNLIKE the RW Aurigae and T Orionis variables, the spectra of the T Tauri stars almost invariably show emission lines and are nearly all of late spectral type (Table XI). A common feature is the large ultraviolet excess that is present on the shortward side of 3500 Å.

Anderson and Kuhi (1969) have investigated several T Tauri variables using both the 120-in. Lick telescope and the 200-in. Palomar reflector, concentrating their investigations mainly on the variable star AS209 (visual range $11^m.3$–$12^m.4$). This particular variable is associated with a patch of bright nebulosity and has a very rich emission spectrum that is representative of most T Tauri stars.

The Balmer lines to H22 and the Paschen lines to P14 are all present in emission together with intense lines of Ca I (H and K), the triplet of Ca II in the infrared, the multiplets of Fe II and Ti II, those of Fe I at 4063 and 4132 Å, and He I at 10830 Å. In addition, the emission spectrum includes weaker lines of Mg I and some other metals, and very weak lines of [O II] at 3727 Å and [S II] at 4068 Å. It may be significant that the forbidden lines of oxygen and sulphur are not found on coudé plates taken during 1967–68.

An absorption spectrum is also present but here it is found that the absorption lines are broad and, in many cases, filled in by continuous emission. From the UBV colours, Anderson and Kuhi made an estimate of the spectral type of AS209 as K5 V, but as these colours were not corrected for reddening it is possible that this spectral type may have to be modified in the light of future work.

Line Broadening in the Spectra of T Tauri Stars

In most of the T Tauri variables that have been investigated spectroscopically, it is found that the emission lines of the higher members of the Balmer series are broadened and tend to blend into what has been termed the "anomalous blue continuum". Two theories have been advanced to explain this effect.

Using a dispersion of 430 Å/mm, Bohm (1957) has shown that the blue continuum begins shortward of 3760 Å and has suggested that this is due to turbulence in the atmospheres of these variables. A turbulent velocity of only 50 km/sec would be quite sufficient to produce this effect. Kuhi (1964) has re-examined the spectroscopic evidence using a dispersion of 16 Å/mm and found that the blue continuum does not begin until 3690 Å. From the width of the H8 emission line, and in the absence of any rotational velocity, it is necessary to postulate a turbulent velocity of approximately 100 km/sec. Since it is almost certain that some rotational velocity is present in these stars, this latter figure is almost certainly somewhat high.

The second hypothesis, put forward by Gordon (1957, 1958), is that synchrotron emission in the infrared from certain well-defined regions on the stellar surface may be responsible for this line broadening. According to this theory, transitions among the upper energy levels are induced by polarized infrared emission and this decreases the mean life time of these transitions, resulting in the observed broadening of the emission lines. A small measure of support for this idea comes from the work of Anderson and Kuhi (1969) on the variable AS209 which shows that not only is an infrared excess present in this star but there is also some correlation between this infrared excess and the various changes in both the blue and ultraviolet continuum. One objection to this hypothesis is that radio emission is usually produced by the synchrotron process. So far, no radio emission has been positively identified as originating in a T Tauri variable.

Taking all of the evidence into consideration, it is probable that both mechanisms are operative in these stars, but the presence of turbulent motions in their atmospheres appears to be a more likely explanation of this effect.

The Ultraviolet Excess

As we have already seen, the spectra of the T Tauri stars generally show an ultraviolet excess shortward of about 3600 Å and, taken as a class, these stars have a fairly well-established relation between this ultraviolet excess and the intensity of the Hα emission lines. This was first demonstrated by Kuhi (1966a). A few members of the group, however, do not show such a relation, notably AS209.

UBVRI photometric observations have been made of several T Tauri stars by Lee (1970) and a comprehensive study of such variables showing a pronounced ultraviolet excess, both in the Orion Nebula and in NGC 2264, has been made by Walker (1966, 1969). Approximately half of the stars examined by Walker show an inverse P Cygni (or YY Orionis) type spectrum. Whereas the normal T Tauri variables have absorption components displaced towards the violet compared with the emission lines, this small subgroup have redward-displaced absorption lines of hydrogen, and, at times, of Ca II.

The emission lines have radial velocities very similar to that of the cluster in which the stars are situated, but the absorption components possess radial velocities that are between 150 and 400 km/sec more positive than the cluster velocity.

For those T Tauri variables with a normal P Cygni spectral type, it is generally believed that this is indicative of ejection of matter from the star and this will be discussed in more detail later in Chapter 20. Here we shall examine the subgroup having a "YY Orionis" type spectrum.

One possible explanation of this anomalous effect is, of course, that of orbital motion in terms of a binary system. However, the spectrograms taken by Walker (1966) show that the absorption lines are always redward displaced with the exception of two spectra obtained of YY Orionis itself. The logical assumption, therefore, is that here we have infall of material onto the stellar surface. Confirmation of this comes from the observation that the "YY Orionis" effect appears only among those variables having an ultraviolet excess and, particularly, among the brighter members of this subgroup.

Further confirmation of this idea would be provided if a correlation could be found between the brightness of the star and the intensity of the inverse P Cygni absorption lines since it appears logical that the luminosity will be affected by the amount of material that is falling onto the star.

Using the prime-focus spectrograph of the 120-in. Lick reflector and a dispersion of

The Nebular Variables

TABLE XIII. SPECTROSCOPIC AND PHOTOMETRIC OBSERVATIONS OF SU ORI

Plate No.	Date (UT)	Exp. (min.)	Radial velocity (km/sec) Emission	Radial velocity (km/sec) Absorption	Magnitude (visual)
ES-384	27 Jan. 1963	165	-40 ± 8	$+332\pm14$	14.5
ES-926	30 Nov. 1964	206	$+ 3\pm2$	—†	15.3
ES-936*	27 Jan. 1965	107	$+ 1\pm13$	—	15.3
ES-950	30 Jan. 1965	330	$+53\pm3$	—†	15.0
ES-953	31 Jan. 1965	285	$+39\pm5$	$+382\pm11$	14.5–15.0
ES-1206	19 Jan. 1966	257	$+17\pm7$	—†	15.6

*Plate underexposed, only emission lines visible.
†Absorption lines absent.

96 Å/mm, Walker (1966) has compared the intensity of the absorption features of SU Orionis with photometric observations of the star taken simultaneously on photovisual plates with the 20-in. Carnegie astrograph. The results are given in Table XIII.

As Walker (1966) has pointed out, the picture is not as straightforward as might have been hoped from these observations. In particular, the wide variations in the radial velocities of the emission lines are worthy of note, especially the negative velocity determined from Plate ES-384. On this occasion, the "YY Orionis" type spectrum was relatively intense and it is possible that the negative radial velocity was due, in part at least, to encroachment of the absorption lines upon the red borders of the emission lines.

Anderson and Kuhi (1969) have examined which of the conventional emission sources (free-bound, free-free and two-photon emission) is responsible for the ultraviolet excesses shown by the majority of the T Tauri variables, again confining their attention mainly to the variable AS209.

Figure 43 illustrates the computed theoretical curves for the free-bound and free-free continuous emission of hydrogen in the regions of the Balmer and Paschen jumps for the special case where the medium is optically thick in the Lyman region.

Figure 44 shows the relation between γ, the energy coefficient, and wavelength for the hydrogen $2s \rightarrow 1s$ two-photon transition for an electron density of $N_e = 10^4$ cm^{-3}, together with the variation of the energy coefficient with density at a temperature $T_e = 10,000°K$ and at wavelengths of 3220, 5100 and 7300 Å.

In computing the curves shown in Fig. 44 it is necessary to introduce a depopulation factor which is dependent upon the $2s \rightarrow 2p$ collisions which also play a part in depopulating the $2s$ level. This factor has values varying between 0.32 and 1.0. In Fig. 44 this has been taken as 0.32.

While it is theoretically possible to make use of the Balmer jump to derive an approximate temperature for a star, this is seldom practical since it is necessary to know the magnitude of the jump relative to zero intensity. Such an ideal case is scarcely ever encountered as very often there is an underlying continuum due to some other source whose intensity is not known.

However, we can plot the electron temperature of hydrogen emission as a function of the fraction of the contribution provided by other sources for different Balmer jumps as in Fig. 45.

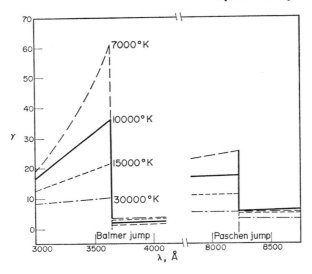

Fig. 43. Computed theoretical curves for the free-bound and free-free continuous emission of hydrogen near the Balmer and Paschen jumps. (After Anderson and Kuhi.)

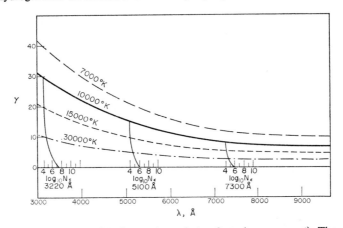

Fig. 44. Continuous emission of hydrogen (two-photon $2s \rightarrow 1s$ component). The units of γ are 10^{-14} cm^3 sec^{-1}. The smaller graphs show the dependence of γ on electron density N_e at three wavelengths for a temperature of 10,000°K. (After Anderson and Kuhi.)

In Fig. 45, I_{Hem} represents the electron temperature of the hydrogen emission and I_s is the fraction of the contribution from other sources. If we now know the curves of various black body temperatures, we may then use Fig. 45 to determine which of these curves is the best fit. Anderson and Kuhi have shown that, for AS209, the black body temperature that best fits the continuum of this star close to 3650 Å is 3500°K. In addition, it appears that the two-photon process does not play any part at all in producing the observed ultraviolet excess and consequently the electron density of the surrounding envelope must be greater than 5×10^5 cm^{-3}.

The Hα Emission Line Profiles

From a study of the Hα emission line profiles in the spectra of T Tauri variables, Kuhi

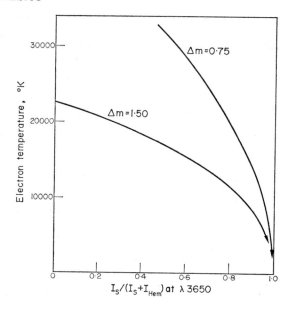

FIG. 45. Electron temperature versus the ratio of I_s to $I_s + I_{Hem}$ at 3650 Å for two Balmer discontinuities. (After Anderson and Kuhi.)

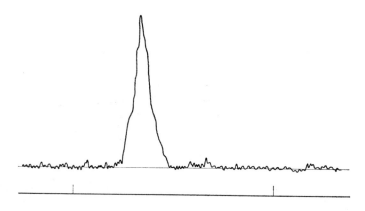

FIG. 46. Hα-line profile for T Tauri. (After Dibaj and Esipov.)

(1964, 1966b) put forward a model for these stars consisting of an expanding mass-loss
envelope.

Dibaj and Esipov (1969) have carried out observations on eight stars which are intimately
connected with nebulosity using an image-tube spectrograph with the 50-in. reflector of the
Sternberg Astronomical Institute. The Hα emission line profiles of two typical T Tauri
variables are given in Figs. 46 and 47.

These are found to be intermediate in character between those observed for objects such
as FU Orionis which possess expanding envelopes together with self-absorption, and "pole-
on" stars such as V380 Orionis with rotating envelopes (Fig. 30).

Quite clearly, the motions within the atmospheres of these variables are not simply those
due to expansion alone.

Lithium Over-abundance

Most of the T Tauri variables that have been examined spectroscopically, show the strong resonance doublet of Li I at 6708 Å which was first observed in the spectra of T Tauri and TY Tauri by Hunger (1957). Determination of the intensity of this doublet indicates that lithium is over-abundant in these stars by at least two orders of magnitude compared with the solar atmosphere. Bonsack and Greenstein (1960) have determined the abundance ratio of Li I 6708/Ca I 6572 in the T Tauri stars as between 4×10^{-4} and 3×10^{-3} while the same ratio for the solar atmosphere as found by Greenstein and Richardson (1951) is only 7×10^{-6}. Other determinations of the lithium abundance in both T Tauri variables and red subgiants have been made by Herbig and Wolff (1966) and Feast (1966).

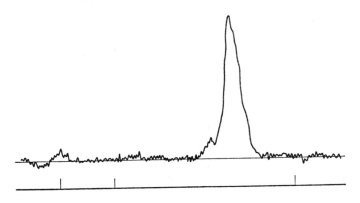

Fig. 47. Hα-line profile for RY Tauri. (After Dibaj and Esipov.)

At this point we must examine the question of whether there is an abnormal over-abundance of lithium in the T Tauri stars or an anomalous depletion of lithium in the Sun. Examination of the spectra of several G- and M-type dwarfs younger than the Sun shows that there is a relation between the amount of lithium and stellar age; stars younger than the Sun having higher concentrations of lithium.

For those stars in which the lithium content has been measured, we also find a curious situation when we compare their positions on the Hertzsprung–Russell diagram. The lithium-rich stars (with only one exception) all lie on the main sequence or in the region of the red subgiants. In the intermediate region, there is a marked increase in the number of low-lithium content stars while the number of lithium-rich stars falls dramatically. From this observation it would seem that there must be a depletion stage in the intermediate region which is then followed by a stage of fresh synthesis, possibly brought about by a renewal of spallation reactions of high energy protons on carbon, nitrogen, oxygen or heavier elements as suggested by Fowler *et al.* (1962).

Spitzer (1949) and Spitzer and Field (1955) have estimated that the abundance ratio of lithium to calcium for the interstellar medium is about 3×10^{-3} which is very similar to that found in the T Tauri variables.

The formation of lithium in stellar interiors where the temperature is of the order of 10^7 °K has been examined by Cameron (1955) and Fowler (1958) who have suggested the following nuclear reactions.

The Nebular Variables

$$He_2{}^3 + He_2{}^4 = Be_4{}^7 + \gamma \tag{1}$$

$$Be_4{}^7 + e^- = Li_3{}^7 + \nu. \tag{2}$$

Since it seems highly probable that near the centre of a star the nuclei of beryllium will be completely broken down, it is likely that the above reactions will take place on the stellar surface in the T Tauri stars.

The fact that, according to measurements made by Babcock (1958), T Tauri has an extremely small magnetic field, makes it difficult to tie in the production of lithium with magnetic activity as suggested by Bonsack (1961) for certain A type stars. In the absence of any further evidence, however, we must conclude that the mechanism of lithium production in the T Tauri stars is similar to that found in the strongly magnetic A type stars examined by Bonsack.

From the work of Andouze, Gradsztajn and Reeves (1967) it is quite clear that any process for the formation of lithium in these variables must involve the participation of enormous amounts of energy stored in the form of high energy particles. For an abundance $A = (N_{Li}/N_H) = 10^{-9}$ in the optical surfaces of these stars, it is calculated that approximately 10 KeV per hydrogen atom of the surface must have been stored in the form of such high energy particles and subsequently lost through electronic collisions.

Peculiarities in the Spectra of T Tauri Variables

Certain peculiarities in the spectra of these stars, for example, the anomalous intensities of the emission lines and variations in the so-called "blue continuum", have been reviewed by Sikorski (1972). It is suggested that many of these abnormalities may be due to regions of atmospheric plasma which are not in thermodynamic equilibrium. The intensity of these various phenomena caused by such regions of plasma is shown to depend, to a large extent, upon the evolutionary stage of the protostar.

Hydroxyl Emission from T Tauri Variables

A large number of infrared objects have been found to be associated with hydroxyl emission at wavelengths 1612, 1665, 1667 and 1720 MHz and these are discussed in Chapter 25. This discovery has prompted several surveys for similar emission from the nebular variables, particularly those of the T Tauri type, which also have intense infrared fluxes. Seven T Tauri variables were examined for OH radio emission during the survey of infrared stars carried out by Wilson and Barrett (1970) which resulted in the detection of the first OH sources to be positively identified with stellar objects. Negative results were obtained for all of the T Tauri variables examined. The wavelength used in this search was 1612 MHz.

From their negative observations, Wilson and Barrett concluded that, in spite of their large infrared excesses, the young T Tauri stars do not provide the necessary conditions for OH-infrared masering which is believed to be the source of this radio emission.

In a later survey by Gahm and Winnberg (1971), the positions of several T Tauri variables in the Auriga and Taurus dark clouds were searched for OH radiation at 1612 and 1667 MHz using the 100-in. radio telescope of the Onsala Space Observatory. In this instance, positive results were obtained for both RY Tauri and SU Aurigae at 1667 MHz. The latter star has been variously classed as an irregular variable and a probable member of the RW Aurigae type.

The hydroxyl velocities of these two variables are in good agreement with results found for the general interstellar field ($\sim +6$ km/sec). Several of the T Tauri variables were also observed at 1612 MHz with negative results up to the upper limits of between 1.2 and 2.3 f.u. of the observed flux density.

Herbig (1970) has proposed the hypothesis that young stars with masses $\sim 1.5\ M_\odot$ may accrete material from the surrounding cloud and subsequently eject material into the immediate environment (~ 15 per cent of the original condensation). This ejected material will contain an appreciable amount of heavier elements, e.g. Mg, Si, Fe and Ti, most of which will be bound up in the form of grains. There will also be diatomic and complex molecules formed from H, C, N and O while the inert gases, He, Ar, etc., will be returned unchanged.

As shown by Heiles (1968) and Cudaback and Heiles (1969), this ejection mechanism can give rise to a very high concentration of interstellar molecules in the dense clouds around these stars.

Hyland *et al.* (1969) have suggested that a close binary system surrounded by a dense circumstellar dust shell can also give rise to 1612, 1665 and 1667 MHz microwave emission lines of OH. Here, the higher velocity component of the 1612 MHz emission has a radial velocity corresponding to the stellar absorption lines and therefore originates in the star. The lower velocity component of this line and the 1665 and 1667 MHz lines are attributed to the dust shell.

Neither of these models can be strictly related to a variable such as RY Tauri since in both cases, the 1612 MHz line should be the most intense and this line is absent in this T Tauri variable. In view of the uniqueness of this observation for RY Tauri and SU Aurigae, further observations are clearly desirable before a suitable mechanism can be put forward.

References

ANDERSON, L. and KUHI, L. V. (1969) *Non-Periodic Phenomena in Variable Stars*, p. 93, Reidel, Dordrecht.
ANDOUZE, J., GRADSZTAJN, E. and REEVES, H. (1967) *Mem. Soc. Roy. Sci. Liége* 15, 299.
BABCOCK, H. W. (1958) *Astrophys. J. Suppl. Ser.* 3, p. 141.
BOHM, H. K. (1957) *Z. Astrophys.* 43, 245.
BONSACK, W. K. and GREENSTEIN, J. L. (1960) *Astrophys. J.* 131, 83.
BONSACK, W. K. (1961) *Ibid.* 133, 551.
CAMERON, A. G. W. (1955) *Ibid.* 121, 144.
CUDABACK, D. and HEILES, C. (1969) *Astrophys. J. Lett.* 155, L21.
DIBAJ, E. A. and ESIPOV, V. F. (1969) *Non-Periodic Phenomena in Variable Stars*, p. 107, Reidel, Dordrecht.
FEAST, M. W. (1966) *Mon. Not. Roy. astr. Soc.* 134, 321.
FOWLER, W. A. (1958) *Astrophys. J.* 127, 551.
FOWLER, W. A., GREENSTEIN, J. L. and HOYLE, F. (1962) *Geophys. J. RAS* 6, 148.
GAHM, G. F. and WINNBERG, A. (1971) *Astron. and Astrophys.* 13, 489.
GREENSTEIN, J. L. and RICHARDSON, R. S. (1951) *Astrophys. J.* 113, 536.
GORDON, I. M. (1957) *Astr. Zh.* 34, 739.
GORDON, I. M. (1958) *Ibid.* 35, 458.
HEILES, C. (1968) *Astrophys. J. Lett.* 151, 919.
HERBIG, G. H. and WOLFF, R. J. (1966) *Ann. Astrophys.* 29, 593.
HERBIG, G. H. (1970) Colloque Liége, Evolution stellaire avant la sequence principale.
HUNGER, K. (1957) *Astron. J.* 62, 294.
HYLAND, A. R., BECKLIN, E. E., NEUGEBAUER, G. and WALLERSTEIN, G. (1969) *Astrophys. J. Lett.* 158, 619.
KUHI, L. V. (1964) *Astrophys. J.* 140, 1465.
KUHI, L. V. (1966a) *Publ. astr. Soc. Pacif.* 78, 430.
KUHI, L. V. (1966b) *Astrophys. J.* 143, 991.
LEE, T. A. (1970) *Publ. astr. Soc. Pacif.* 82, 765.

The Nebular Variables

SIKORSKI, J. (1972) *Postepy Astron.* (Poland) **20**, 37.
SPITZER, L. JNR. (1949) *Astrophys. J.* **109**, 548.
SPITZER, L. JNR. and FIELD, G. B. (1955) *Ibid.* **121**, 300.
WALKER, M. F. (1966) *Stellar Evolution*, p. 405, Plenum Press, New York.
WALKER, M. F. (1969) *Non-Periodic Phenomena in Variable Stars*, p. 103, Reidel, Dordrecht.
WILSON, W. J. and BARRETT, A. H. (1970) *Astrophys. J.* **160**, 545.

CHAPTER 16

Physical characteristics of T Tauri stars

THE VARIABLE stars of the T Tauri class are all dwarfs with spectra of types G, K or M. Very few are known which do not show emission lines in their spectra and in this respect they differ from the variables of the two preceding classes.

The majority of the T Tauri stars show an excess infrared emission like other nebular variables but in this case it is not dominant. Consequently, we may interpret these variables as stars, newly formed out of their parent protostellar clouds, in which the last remnants of these clouds have almost disappeared. The calculations made by Larson (1972a) indicate that for a relatively short period after the stellar core becomes visible there should still be some infall of material onto the star (observable in the case of the YY Orionis stars). Approximately 10^5 years later, this infalling material will either be totally accreted by the star or removed by mass ejection.

On the Hertzsprung–Russell diagram, these variables appear above the main sequence. There is, however, quite a wide scatter of positions in spite of the fact that their masses are, as we shall see, fairly uniform.

As Larson (1972a) has pointed out, this scatter may be due to variations in the amount of dusty material still surrounding the stars or differences in both the initial and boundary conditions of the protostellar cloud.

The existence of circumstellar shells of dust will give rise to at least part of the observed scatter as shown by Strom *et al.* (1971, 1972a, 1972b) and Breger (1972).

The physical characteristics, too, will depend upon such factors as the initial temperature of the protostellar cloud, its radius, mass and density. Rotational effects and the presence of magnetic fields may also have quite pronounced effects. Larson (1972a) has shown that, on the basis of non-homologous collapse of the protostellar cloud, a change in the initial temperature by a factor of 2 will correspond to a factor of 2 in the stellar radius, one of 8 in the density and a factor of 2.8 in the collapse time, on the assumption that the Jeans criterion is satisfied. Similar variations in the end result are found, for example, by a reduction in the radius of the initial cloud by a factor of 2.

For a star of a given mass, any change in the collapse time will result in the stellar core becoming visible in the optical region at different stages of its pre-main sequence evolution.

As far as the boundary conditions are concerned, it is not an easy matter to define such conditions precisely. Whether a definite boundary exists at all for a collapsing protostar is, perhaps, doubtful. During the various stages of its pre-main sequence evolution, such an object could accrete material that did not originally form a part of the parent protostellar cloud. In this instance, therefore, its mass as it reaches the main sequences will be larger than that indicated by the Jeans criterion.

The Nebular Variables

Masses of the T Tauri Variables

Kuhi (1966) and Herbig (1967) have reviewed the properties of typical T Tauri variables. Much of the evidence suggests that these stars have masses in the range 0.5–1.5 $M_\odot$. This is in good agreement with their spectra which, in spite of the superimposed peculiarities due to the surrounding dust clouds, are those of dwarf stars. Any enhanced lines are normally very weak.

THEORETICAL RESULTS

Both the homologous collapse theory examined by Hayashi (1961, 1970) and Hayashi *et al.* (1962), and that based upon non-homologous collapse suggested by Larson (1969, 1972a) yield results which agree closely for low mass stars in the final stages of their pre-main sequence evolution. For objects in the mass range found for T Tauri variables, much of their mass is largely convective in nature.

As a result, both theories show that such stars will first appear near the ends of their respective "Hayashi" tracks with total masses very similar to those actually observed. As we shall now see, this also applies to other physical characteristics of these variables.

Effective Temperatures of the T Tauri Variables

Herbig (1962) has estimated the effective temperatures of a number of T Tauri stars and shown that these lie between 5.5×10^3 and $8.5 \times 10^{3\circ}$K. From Table V it will be seen that these values coincide with the theoretical properties of newly formed stars having masses from 0.25 to $\leqslant 1.5\,M_\odot$, e.g. for $3.57 < T_e < 3.67$.

It will be noted that the effective temperatures given above are somewhat higher than those normally quoted for stars of spectral types G–M in the Draper Catalogue. The latter apply, in general, to main sequence stars whereas the nebular variables lie above the main sequence.

THE SURROUNDING CIRCUMSTELLAR SHELLS

Although the circumstellar shells around the majority of T Tauri variables appear, from the smaller excess infrared emission, to be thinner than those around such objects as R Monocerotis and R Coronae Australis, their temperatures lie in the same range from 600 to 900°K.

Luminosities of the T Tauri Variables

Local obscuration is not such a dominant factor here as it is in the case of the RW Aurigae and T Orionis variables and reasonably accurate estimates of the luminosities of these stars have been made. The observations of different authors are in fairly good agreement, yielding values of 0.8–3.5 $L_\odot$. Reference to Table V shows that, again, these figures agree closely with the properties of newly formed stars in the mass range we have been considering.

Quite clearly, the T Tauri variables lie at the lower end of the luminosity range for the nebular variables.

Stellar Radii

According to Herbig (1962, 1967) most of these stars have radii between 1 and 3 $R_\odot$ with a peak at $\sim 2\, R_\odot$. These small radii are more in agreement with the non-homologous collapse models than those based upon the homologous collapse of the protostellar cloud which predict radii of the order of 60 $R_\odot$ for a star of 1 $M_\odot$, by the time the star comes onto its "Hayashi" track, calculated by Ezer and Cameron (1965).

Binary Systems among T Tauri Variables

Most of the theories of stellar evolution which take rotation of the protostellar cloud into account, for example Larson (1972b), Moss (1973) and Porfiriev *et al.* (1969), result in the formation of multiple systems due to fragmentation. The simplest possibility and also that which represents the lowest order deviation from axial asymmetry, is fragmentation into a binary system. Even in the case of multiple systems, the *n*-body calculations of Agekyan and Anosova (1968) indicate that, owing to the inherent instability of these systems, ejection of stars will take place until a stable binary is left. On this basis, we would expect to find such systems among the T Tauri variables and these have actually been observed in significant numbers. The characteristics of typical binaries are given below.

UZ Tauri

This variable has been examined by Joy (1942). The Mount Wilson photographs show that a faint red double star lies in the position of this object. The components have visual magnitudes of $13^m.0$ and $13^m.3$ respectively, the separation being $3''.5$. Both spectra are of early dMe type showing intense emission lines of the Balmer series and Ca II. Both of these stars are variable with amplitudes of $\sim 1^m.5$.

CZ and DD Tauri

As mentioned earlier in Chapter 14, this pair of T Tauri variables have a separation of $31''$ and spectral types of dM2e and dK6e respectively. It seems almost certain that these stars form a binary system although no figures for their orbital elements or period of revolution have been published. The light curves of these two stars are given in Fig. 39.

DH and DI Tauri

This faint T Tauri binary system has a separation of $16''$ and, as before, it appears to be a physically connected system although few observational details are available. The spectral types of the components are both dM0e.

RS Canum Venaticorum Binaries

This small subgroup of the Algol eclipsing binaries is characterized by having variable periods and one component which is apparently a typical dwarf. The prototype star has been described by Sitterley (1931). The visual range is $8^m.00$–$9^m.27$ with a secondary minimum of $8^m.03$. The spectral types are F4n and dG8 and $P = 4^d.797875 + 0^d.027 \sin 0°.18E$.

The evolutionary status of these variables has recently been investigated by Hall (1972) who has considered three different interpretations for their evolution, namely post-main sequence before and after mass exchange and pre-main sequence. The last of these is the

111

The Nebular Variables

most successful in explaining the observed properties of these binaries and is the one preferred.

An important aspect of this interpretation is that the cool, dwarf component, particularly that in the case of RS Canum Venaticorum itself, displays many of the properties of the T Tauri variables. If this is the correct explanation, it may throw new light on the various pre-main sequence evolutionary stages of the T Tauri stars.

The Circumstellar Envelopes around T Tauri Variables

As may be expected from their evolutionary state, there are fairly wide variations observed in many of the properties of the circumstellar envelopes observed around these stars. Such parameters as density, opacity and total amount of material present vary from star to star.

One of the most fully investigated properties of these envelopes is the nature and formation of the grains which form a not insignificant percentage of their composition. In recent years, several authors have reported on the properties of these grains and the possibility of nucleation in these circumstellar shells.

Snow and Wallerstein (1972) have examined the spectra of a number of stars, including some T Tauri variables, which show infrared excesses and other features indicative of dust-containing envelopes. In particular, a search has been made for diffuse features in their spectra at wavelengths of 4430, 5780, 5796 and 6613 Å. In no case were such diffuse bands positively identified that could be ascribed to the circumstellar envelopes and it was concluded that the grains present around these stars cannot be identical with those found in the general interstellar medium.

Svatos (1972) has examined the problem of estimating the scattering properties of rough grains in these shells by means of determinations in the far infrared. His results indicate that these properties may be obtained far more accurately and conveniently from the intensity of the scattered light than from the extinction cross-section of such grains.

Nucleation in Expanding Envelopes

During its final evolution onto the main sequence when most, or all, of the surrounding dust cloud has been accreted, mass ejection may take place resulting in an expanding envelope around the star. Evidence for this is found in the Doppler-shifted absorption components on the longward borders of the emission lines.

The possibility that, with a lowering of the temperature in such expanding envelopes, nucleation of dust grains may occur, has been examined by Nishida and Nakazawa (1973). Assuming that only graphite will condense under these conditions, the nucleation time scale I_n has been investigated and compared with the expansion time scale I_{exp} on the assumption that in the gas ejected from a T Tauri variable the flow is steady and spherically symmetric.

If the concentration of carbon in the gas, by mass, is 10^{-2} and the gas expands adiabatically, the mass loss rate required for nucleation is found to be larger than those which have been observed in the T Tauri stars. On this basis, it appears that nucleation around these variables will be difficult, if not impossible.

Mass Loss in T Tauri Variables

Apart from those variables with the "YY Orionis" type spectrum where accretion of material is still taking place, the majority of these stars show evidence of ejection of mass into a surrounding envelope. The explanation of these two effects appears to be one of stellar age, only the youngest of these variables exhibiting evidence of accretion. As their evolution proceeds, a point will be reached where the total mass of the protostellar cloud has been accreted by the star or the remaining remnants have been disrupted and dispersed by radiation pressure. Following this, mass ejection begins.

The dynamics of this mass loss have been examined by Gilman (1972) on the basis of coupling of the grains to the gas in the circumstellar shells. While the dust grains are not position-coupled, they are momentum-coupled to the gas and therefore radiation pressure on the grains can drive the mass loss mechanism.

References

AGEKYAN, T. A. and ANOSOVA, ZH. P. (1968) *Astrofizika* **4**, 31.
BREGER, M. (1972) *Astrophys. J.* **171**, 539.
EZER, D. and CAMERON, A. G. W. (1965) *Can. J. Phys.* **43**, 1487.
GILMAN, R. C. (1972) *Astrophys. J.* **178**, 473.
HALL, D. S. (1972) *Publ. astr. Soc. Pacif.* **84**, 323.
HAYASHI, C. (1961) *Publ. astr. Soc. Japan* **15**, 450.
HAYASHI, C., HOSHI, R. and SUGIMOTO, D. (1962) *Prog. Theor. Phys.*, Suppl. No. 22.
HAYASHI, C. (1970) *Evolution Stellaire Avant la Sequence Pricipale* p. 127, Univ. of Liége.
HERBIG, G. H. (1962) *Adv. Astr. Astrophys.* **1**, 47.
HERBIG, G. H. (1967) *Scient. Am.* **217**, 30.
JOY, A. H. (1942) *Publ. astr. Soc. Pacif.* **54**, 33.
KUHI, L. V. (1966) *J. R. astr. Soc. Canada* **60**, 1.
LARSON, R. B. (1969) *Mon. Not. Roy. astr. Soc.* **145**, 271.
LARSON, R. B. (1972a) *Ibid.* **157**, 121.
LARSON, R. B. (1972b) *Ibid.* **156**, 437.
MOSS, D. L. (1973) *Ibid.* **161**, 225.
NISHIDA, S. and NAKAZAWA, K. (1973) *Prog. Theor. Phys. Japan*, **49**, 1152.
PORFIRIEV, V. V., SHULMAN, L. M. and ZHILYAEV, B. E. (1969) *Nature* **222**, 255.
SITTERLEY, B. W. (1931) *Princ. Contr.*, No. 11.
SNOW, T. P. JNR. and WALLERSTEIN, G. (1972) *Publ. astr. Soc. Pacif.* **84**, 492.
STROM, K. M., STROM, S. E. and YOST, J. (1971) *Astrophys. J.* **165**, 479.
STROM, S. E., STROM, K. M., BROOKE, A. L., BREGMAN, J. and YOST, J. (1972a) *Ibid.* **171**, 267.
STROM, S. E., STROM, K. M., YOST, J., CARRASCO, L. and GRASDALEN, G. (1972b) *Ibid.* **173**, 353.
SVATOS, J. (1972) *Astrophys. & Space Sci.* **17**, 238.

CHAPTER 17

Spatial distribution

THROUGHOUT the preceding chapters it has been made clear that the distinction among the RW Aurigae, T Orionis and T Tauri variables cannot yet be precisely defined, although the T Tauri stars can be somewhat more readily differentiated by virtue of their characteristic spectra and light variations. In the past, however, there has inevitably been some uncertainty regarding classification, especially in the case of the fainter stars.

Overall Distribution

There is now ample evidence, both observational and theoretical, to suggest that the T Tauri variables are very young, gravitationally contracting objects either just on the main sequence or still evolving towards it. Consequently, we would expect to find these variables in fairly large numbers in regions where star formation is possible and currently going on, i.e. in areas of dark and bright nebulosity.

Large numbers of these stars are found in dense aggregates, the T-associations, first defined by Kholopov (1951) in the area around ε Chamaeleontis and now extended to similar regions in Auriga, Perseus, Taurus and Orion in particular.

Herbig (1962) has estimated that there are approximately 12,000 T Tauri stars situated within 1 kpc of the Sun, while Kuhi (1964) has calculated a total of $\sim 10^6$ within the Galaxy. The majority appear to lie in anticentre zones, in or near large H II regions (Table XIV). Ideal regions would seem to be areas where heavy obscuration is present, indicative of concentrations of dust grains, although this is not always the case. We find very few, for example, in the area of the Aquila Rift in the Milky Way.

TABLE XIV. GALACTIC COORDINATES OF T TAURI ASSOCIATIONS

Constellation	Galactic coordinates*	
	l^{II}	b^{II}
Cygnus	46°	+ 1°
Taurus	144°	−11°
Taurus	145°	−19°
Monoceros	171°	+ 3°
Orion	178°	−20°
Chamaeleon	269°	−17°

*Epoch 1950.0.

The T Tauri variables appear to be generally absent from regions in which OB-type stars and Wolf-Rayet stars are present and, curiously, none have been discovered in the prominent H II region of NGC 2237/2244 (the Rosette Nebula) or around the Bok globules which are small, compact and sharply-defined dark clouds visible against the brighter background of certain gaseous nebulae.

In general, we find the T Tauri variables situated in a fairly wide band paralleling the galactic plane, extending from Cygnus to Monoceros with some members as far south as the T-association in Chamaeleon. The main concentration of these stars (and the closely allied Hα-emission objects) appears to be in the large nebulous regions of Taurus and Orion.

Distribution of T Tauri Variables in Specific Regions

THE CYGNUS REGION

A survey for Hα-emission stars in the region around γ Cygni has been made by Kazaryan and Parsamian (1971) using a 40-in. Schmidt telescope with a 4° objective prism. A total of 35 new stars of this type, having short spectrum and showing the Hα line in emission have been discovered in this area. While the majority of these appear to be constant in brightness, some are almost certainly T Tauri variables.

Jarecka (1972) has carried out a photographic survey in the near infrared of stars in the Cygnus OB2 association. The spectra of 5 of these stars show strong absorption bands of TiO and these are clearly cool M-type giants. The positions of a further 15 stars on a (I/V–I) Hertzsprung–Russell diagram suggest that they fall upon a reddened pre-main sequence branch in the association and these are probably typical T Tauri variables.

THE TAURUS REGION

As may be seen from Table XIV, a considerable number of T Tauri variables are to be found in the dark clouds in this constellation. The majority are faint objects even at maximum brightness and all have spectral types of dGe to dMe.

A number of these variables have been described by Joy (1945). The variable BP Tauri described by Leman-Balanovskaya (1935) as a short period variable ($P = 0^d.1848$) has since been shown to be a T Tauri variable.

THE MONOCEROS REGION

Some 70 T Tauri variables have been discovered in the neighbourhood of NGC 2264 near S Monocerotis by Herbig (1954) and Wenzel (1955a, 1955b). The majority of these are extremely faint objects whose light curves have not been characterized and it is probable that several are RW Aurigae variables.

Nandy and Pratt (1972) have studied the variability of several T Tauri stars in NGC 2264 in UBVRI colours. It is found that the range of variability in amplitude I is appreciably less than in UB and V. The relative opacities at these wavelengths have been determined from the variations in the different colours of these dust-embedded stars.

THE ORION REGION

Several regions in, and near, the Orion Nebula have been found to contain associations of T Tauri stars in close proximity to the allied T Orionis variables.

Sandeleak (1971) has examined six very faint stars near the Orion Nebula which had not previously been confirmed as members of this class. From a series of objective prism plates, all of these stars were found to show the typical T Tauri type of spectrum with emission lines clearly visible at low dispersion.

NGC 2068 AND 2271

These two regions of reflection nebulosity are situated in a large, dark cloud with dimensions $\sim 2°$ by $4°$ in the Orion Nebula. Along its western boundary, there is a complex of emission and reflection nebulosities lying close to ζ Orionis, while to the north and east lies Barnard's Loop. The latter feature may be an expanding shell formed by a supernova as suggested by Öpik (1953). An alternative theory is that this is a region of gas which has been accelerated by virtue of the presence of hot, young stars which have been formed near its centre as postulated by Savedoff (1956) and Menon (1958). The latter authors have also put forward the hypothesis that star formation is taking place between the shell and the external, undisturbed region where a compressed interface would be expected to exist. While this idea is undoubtedly attractive, it does not fully accord with the observations of Herbig and Kuhi (1963) on the distribution of T Tauri variables in this region.

A total of 45 Hα emission line stars have been discovered in the vicinity of NGC 2068 and 2071 by Herbig (1954). Nine of these had been found during an earlier survey by Haro and Moreno (1953) and most of these were subsequently discovered to be variable by Fedorovitch (1960) and Rosino (1960). These stars, like those found later by Herbig (1954) are almost certainly all T Tauri variables.

The photographic magnitude of the T Tauri stars in this field begins at $\sim 15^m.0$, with the numbers increasing to $m_{pg} = 18.0$, after which there is a fairly steep drop in numbers. As Herbig and Kuhi have pointed out, this may be no more than a selection effect in that for those stars fainter than $18^m.0$, only the members with exceptionally intense Hα appear on the plates and are capable of identification. It is noteworthy, too, that the emission stars in the Orion Nebula, which is at approximately the same distance as NGC 2068, first show themselves in comparable numbers at an apparent photographic magnitude between $2^m.0$ and $2^m.5$ brighter than in NGC 2068. Again, we must not read too much into this effect which may simply reflect the relative richness of the Orion stars in the former nebula.

From the observations of Herbig and Kuhi it is evident that the distribution of these T Tauri variables over this dark cloud is far from uniform. There appear to be two concentrations; one around NGC 2068 and 2071 and another, admittedly more scattered group along the cloud, leading towards a further concentration in the direction of ζ Orionis.

As opposed to these reasonably compact groupings of the T Tauri stars, other regions of the cloud are found in which such variables are absent. One common factor for the presence of these stars appears to be the vicinity of B-type stars. It may be that, due to the configuration of the cloud, these regions of T Tauri variables are more transparent than the others; certainly it is very doubtful if the obscuration can have been removed by the B-type stars themselves which are not sufficiently luminous to reduce the opacity of the dust to this extent. A further possibility, suggested by Herbig and Kuhi, is that conditions in these particular areas may be, somehow, more favourable for star formation.

The hypothesis that star formation was triggered off by compression in the region between the expanding cloud (represented by Barnard's Loop) and the external dark cloud fails to explain the existence of a number of T Tauri variables more than $1°$ outside the Loop in the vicinity of BD+$1°$ 1156.

Recently, a radio survey in the region of NGC 2068 has been carried out by Sancisi (1970) using a 25-m telescope. The H I distribution in this area shows a maximum of peak brightness and a minimum of velocity width very close to the group of T Tauri variables.

Dust in Barnard's Loop

The presence of a large number of T Tauri variables in the vicinity of Barnard's Loop, together with several T Orionis variables of early spectral type (all of which possess extinction curves which differ significantly from the mean interstellar extinction law) implies the existence of dust grains in large quantities. Marked deviations from the usual $1/\lambda$ extinction law in the optical region were first noted by Baade and Minkowski (1937), followed by further confirmatory determinations by Stebbins and Whitford (1945), Borgman (1961), Johnson (1968) and Divan (1971).

Walker (1969) has demonstrated that the mean interstellar extinction is only $\sim 0^{m}.2$ over a distance of ~ 470 pc (the distance to the Trapezium) measured in the same galactic direction as the Trapezium stars. For the T Orionis and T Tauri variables in this region, however, it amounts to $> 2^{m}$. Quite clearly, therefore, the greater proportion of the observed extinction must arise in the medium within the nebula.

Van der Hulst (1949) and Krishna Swamy (1965) have argued that this extinction may be caused by ice grains. For several reasons, this view is not now generally accepted. From theoretical considerations, the stability of such volatile grains in H II regions seems unlikely as shown by Wickramasinghe and Williams (1968), while detailed extinction curves in the optical region obtained by Divan (1971), together with ultraviolet extinction data gained from rocket and orbiting satellite experiments in the ranges 1030–1180 and 1230–1350 Å by Carruthers (1970) and Bless and Savage (1970) suggest dust rather than ice grains.

We also have the evidence of O'Dell and Hubbard (1965) and O'Dell et al. (1965) which has established that dust is present in H II regions. From their results, an estimate may be made of the gas to dust ratio

$$\delta = N_{H}/N_{d}C_{sca} \tag{1}$$

where N_{H} and N_{d} are the respective densities of hydrogen and dust particles and C_{sca} is the mean scattering cross-section of the dust grains. The values of δ obtained by O'Dell et al. are 5×10^{20} cm^{-2} for the outer regions of the Orion Nebula and 144×10^{20} cm^{-2} for the inner regions. The value for a normal H I field is 20×10^{20} cm^{-2}.

Barnard's Loop encloses a region of space which contains, apart from numerous nebular variables of the T Orionis and T Tauri type, a large number of bright O and B type stars, all of which are appreciably reddened. Until it became possible to take ultraviolet photographs of this region, the Loop Nebula was known only as an elliptical feature with major and minor axes of about 110 and 78 pc.

The 2200–4900 Å ultraviolet objective prism photographs taken by the Apollo 11 astronauts, reported by Henize et al. (1967) show it to be a very large structure which is exceptionally bright in this spectral region. However, the luminosity is appreciably smaller in the 1230–2100 Å band since the Loop does not appear on photographs taken by Henry and Carruthers (1970) with a Schmidt camera on a sounding rocket.

O'Dell et al. (1967) have suggested that the excess brightness in the 2200–4900 Å region is due to scattering of ultraviolet stellar radiation by dust in the nebula and the extinction and scattering by such dust grains have been compared with theoretical predictions for mixtures of graphite, iron and silicate grains by Nandy and Wickramasinghe (1971). From

117

their data, these authors conclude that there is a deficiency of iron and graphite grains in this region compared with silicate particles. In addition, the average radius of the silicate grains which agrees best with the infrared, optical and far ultraviolet data is significantly larger than similar grains in normal H I clouds.

In this context, however, there still remains the problem of how such grains can remain stable in H II regions where both optical and ultraviolet radiation pressure from the bright O and B type stars should disperse them quite rapidly. Certainly there will be some collisions between the grains and neutral gas which will provide a braking force, but this is unlikely to be sufficient to extend the short time scale to any appreciable extent.

Mathews (1967) and Wickramasinghe (1970) have examined this problem in some detail and shown that the photoelectric effect due to the vicinity of early type stars will give these grains a positive charge and through Coulomb interactions they will become coupled with the ambient H II gas. The smaller grains will, of course, disperse on the short time scale but larger ones with radii $\geqslant 10^{-5}$ cm can remain stable within an H II region for more than 10^7 years. Such a period embraces the total formation time of the stellar core of a nebular variable.

THE CHAMAELEON REGION

A small number of the nebular variables found in the T-association near ϵ Chamaeleontis are known to be T Tauri stars although many of them belong to the related RW Aurigae type, e.g. T Chamaeleontis. It also appears probable that some T Tauri variables are present among the Hα-emission line objects reported by Mendoza (1972) in the dark region lying to the north of the T-association.

T TAURI VARIABLES IN OH CLOUDS

Grasdalen and Kuhi (1973) have made an extensive survey of eleven dust clouds exhibiting OH microwave emission lines for the presence of Hα-emission stars. In four of these clouds, a total of eight such objects have been discovered which, from their spectroscopic characteristics, appear to be T Tauri variables. None of these stars were found in the remaining seven clouds down to a visual magnitude of $\sim 18^m$. There is, as yet, little evidence that these variables themselves emit radio radiation at the OH wavelengths.

References

BAADE, W. and MINKOWSKI, R. (1937) *Astrophys. J.* **86**, 123.
BLESS, R. C. and SAVAGE, B. D. (1970) *Ultraviolet Stellar Spectra and Related Ground Based Observations*, Reidel, Dordrecht.
BORGMAN, J. (1961) *Bull. astr. Inst. Netherlands* **16**, 99.
CARRUTHERS, G. R. (1970) *Ultraviolet Stellar Spectra and Related Ground Based Observations*, Reidel, Dordrecht.
DIVAN, L. (1971) *Astron. and Astrophys.* **14**, 198.
FEDOROVITCH, V. P. (1960) *Peremennye Zvezdy* **13**, 166.
GRASDALEN, G. L. and KUHI, L. V. (1973) *Publ. astr. Soc. Pacif.* **85**, 193.
HARO, G., MORENO, N. (1953) *Astrophys. J.* **116**, 438.
HENIZE, K. G., WACKERLING, L. R. and O'CALLAGHAM, F. G. (1967) *Science* **155**, 1407.
HENRY, R. C. and CARRUTHERS, G. R. (1970) *Ibid.* **170**, 527.
HERBIG, G. H. (1954) *Astrophys. J.* **119**, 317.
HERBIG, G. H. (1962) *Contr. Lick Obs.* No. 282.
HERBIG, G. H. and KUHI, L. V. (1963) *Astrophys. J.* **137**, 398.
JARECKA, P. J. (1972) *Bull. Am. astr. Soc.* **4**, 233.
JOHNSON, H. L. (1968) *Stars and Stellar Systems* **7**, 221.
JOY, A. H. (1945) *Astrophys. J.* **102**, 168.

KAZARYAN, M. A. and PARSAMINA, E. S. (1971) *Astrofizika* **7**, 671.
KHOLOPOV, P. N. (1951) *Bull. Variable Stars* **8**, 83.
KRISHNA SWAMY, K. S. (1965) *Publ. astr. Soc. Pacif.* **77**, 164.
KUHI, L. V. (1964) *Astrophys. J.* **140**, 1409.
LEMAN-BALANOVSKAYA, I. H. (1935) *Pulkova Circ.* No. 16.
MATHEWS, W. G. (1967) *Astrophys. J.* **147**, 965.
MENDOZA, E. E. (1972) *Publ. astr. Soc. Pacif.* **84**, 641.
MENON, T. K. (1958) *Astrophys. J.* **127**, 28.
NANDY, K. and WICKRAMASINGHE, N. C. (1971) *Mon. Not. Roy. astr. Soc.* **154**, 255.
NANDY, K. and PRATT, N. (1972) *Astrophys. and Space Sci.* **19**, 219.
O'DELL, C. R. and HUBBARD, W. B. (1965) *Astrophys. J.* **142**, 591.
O'DELL, C. R., HUBBARD, W. B. and PEIMBERT, M. (1965) *Ibid.* **143**, 743.
O'DELL, C. R., YORK, D. G. and HENIZE, K. G. (1967) *Ibid.* **150**, 835.
ÖPIK, E. (1953) *Astron. J.* **2**, 219.
REDDISH, V. C. (1967) *Mon. Not. Roy. astr. Soc.* **135**, 251.
ROSINO, L. (1960) *Asiago Contr.* No. 109.
SANCISI, R. (1970) *Mem. Soc. Roy. Sci. Liége* **19**, 313.
SANDELEUK, N. (1971) *Publ. astr. Soc. Pacif.* **83**, 95.
SAVEDOFF, M. (1956) *Astrophys. J.* **124**, 533.
STEBBINS, J. and WHITFORD, A. E. (1945) *Ibid.* **102**, 273.
VAN DER HULST, H. C. (1949) *Rech. astr. Obs. Utrecht* **11**, part 2.
WALKER, M. F. (1969) *Astrophys. J.* **155**, 447.
WENZEL, W. (1955a) *Mitt. veränderl. Sterne*, No. 190.
WENZEL, W. (1955b) *Ibid.* No. 193.
WICKRAMASINGHE, N. C. and WILLIAMS, D. A. (1968) *Observatory* **88**, 272.
WICKRAMASINGHE, N. (1970) *Nature*, **225**, 145.

CHAPTER 18

Interaction with nebulosity

ALL OF the T Tauri stars are associated, in some way, with nebulosity. The available evidence, too, as shown by Herbig (1962), is that these variables are in the pre-main sequence stage of their evolution and contracting out of the gas and dust clouds in which they are immersed. The brightness of these objects at long wavelengths in the infrared suggests that much of the radiation is produced by absorption of visible and ultraviolet radiation from the central stellar core followed by thermal re-radiation at these longer wavelengths from surrounding dust grains.

In many instances we find that the T Tauri stars are in association with cometary nebulae which, like the stars themselves, vary in brightness and also in structure with time. The fact that, in nearly every case, the star is found to occupy a geometrically significant position with respect to the nebula (usually at the tip, less often in the centre as for T Tauri itself) implies that both components are physically connected.

Here we must, therefore, examine the very relevant question of which is the cause and which is the effect. Theoretically, of course, the star may have been formed as a consequence of the presence of the allied nebulosity, or vice versa.

Some early observational work by Minkowski (1942) suggested that both the formation of the nebulae and their luminosities are due to the associated stars. A similar conclusion was reached by Ambartsumian (1955) from a study of the strong blue continuum emission found in many of the spectra of the T Tauri variables. While excluding simple reflection as the cause of the light variations of the nebulae, the latter reasoned that the stars are the basic sources of energy and the nebulae are formed by outflow of material from the embedded stars.

From the work of Kuhi (1964) on mass loss from these young stars it seems plausible that this may result in the formation of a circumstellar cloud, much of which is almost certainly particulate in nature, possibly graphite and silicate grains (thereby explaining the large infrared excesses). It is even possible that there may be a disc-like distribution of cool matter surrounding these stars which are, in general, rapidly rotating objects. Such a disc will extend radially from the equatorial regions of the star.

Other workers who have supported the theory that simple reflection is not the cause of the illumination of the surrounding nebulae are Khachikyan and Parsamian (1965), based upon observation of the biconical nebula around the emission object LkHα-208. Haro (1953) has also examined the difference between both brightness and colour of the cometary nebula Haro 13a and its associated star, concluding that reflection processes cannot explain these differences and that the star has condensed from the original nebulous material.

In contrast to the above, however, Herbig (1956, 1971) has shown that the spectra of both LkHα-101 and its surrounding nebulosity have Hα emission lines of similar intensities and suggests that the nebula is a bright reflection nebulosity. While it is quite true that this star appears to be much too faint to illuminate the nebula (NGC 1579), this could be because of very heavy obscuration since the star is situated within a dark dust lane that lies across the nebula.

We may here cite the cases of FU Orionis and LkHα-190, both of which have brightened appreciably over the past 30 or 40 years. Shortly after these increases in brightness, small nebulous arcs were observed in close proximity to the stars. Herbig (1966) and Herbig and Harlan (1971) have postulated that here we are witnessing the propagation of light between the stars and dust clouds already in existence. Quite clearly, there is some evidence for reflection effects in these objects.

There are, of course, several other theories which have been advanced from time to time to explain, in particular, the small diffuse nebulae associated with many T Tauri variables and early type emission stars. Since we find the T Tauri variables in large numbers within regions of dark and bright nebulosity, the chance of a luminous star entering a region of cold gas is quite high. On this basis, Dibaj (1960, 1963) has postulated that ionizing radiation from such a star may result in the focusing of shock waves whenever it meets such a cloud of cool gas.

Poveda (1965a), on the other hand, has suggested that the observed nebulae are rim structures and, this being so, the material within such structures will be at a comparatively high density. Conditions are therefore suitable for gravitational collapse leading to the eventual formation of stars by compression.

Earlier in the chapter, we saw how it is possible that much of the material surrounding a protostar during the pre-main sequence evolutionary phase may be concentrated in a disc-shaped structure situated in the equatorial plane of the protostellar cloud. This idea has been modified and extended by Poveda (1965b) and Herbig (1968, 1971) to explain the appearance of the biconical nebulae around stars such as LkHα-208 whose geometry is indicative of such a disc of dark matter. Although the light from the equatorial regions will be essentially obscured, that from the poles will be scattered by the surrounding medium, giving rise to the biconical shape observed for these nebulae.

Light Variation of Star and Nebula

As we have seen earlier, in Chapters 6 and 12, several of the nebulae associated with these variables are themselves variable in brightness. Evidence of any direct relation between the brightness of the variable and that of the surrounding nebulosity appears to be very slender at the moment. Indeed, there is some evidence that the two phenomena may be either completely independent, or there is an inverse relation between the two.

Herbig (1961) has demonstrated that, when near minimum brightness, the spectrum of RY Tauri exhibits strong emission lines of [S II] at 4068 and 4076 Å. According to Joy (1945), these lines are not seen when the variable is near maximum light. These observations support earlier suggestions by Herbig (1953) that the [S II] and [O I] lines in the spectra of the T Tauri variables originate in a relatively high excitation, low density envelope that lies very close to the star. The comparative stability of these lines (i.e. their relative prominence near minimum light) means that this nebulous envelope is scarcely affected by the activity in the star itself which is responsible for the light maxima.

T TAURI AND NGC 1554, 1555

Hind's variable nebula, NGC 1555, was apparently bright in 1852 when first discovered but had faded appreciably when searched for in 1861 by d'Arrest, being visible only in large instruments. Seven years later, it was stated by the same observer to be invisible.

An account of the history of the nebulae near T Tauri has been given by Herbig (1953). Additional information with references can also be found in Herbig (1950).

Large-scale direct photographs taken by Herbig (1950) show a region protruding from the stellar image where the forbidden emission lines of [O I] and [S II] originate. A similar region near RY Tauri has been looked for by Herbig (1961) using the 120-in. Lick telescope and when the star was close to minimum brightness, but without success.

RY TAURI AND B214

RY Tauri apparently illuminates the southern tip of the small elliptical dark cloud Barnard 214. The dimensions of this cloud are approximately $7' \times 11'$. The small, bright nebulosity near RY Tauri was described by Barnard (1907) as composed of two comet-like tails, each about 6' in length, extending from the stellar image in position angles 60° to 330°.

Larger-scale photographs show that this nebulosity is brightest some 15″ or 20″ west of the star and that it extends north from this point, forming a diffuse streamer that gradually fades and turns westward, finally disappearing some 5' from the star. A further feature shown by these plates is that the cometary appendage in position angle 60° is, in reality, a detached wisp of nebulosity.

On two unfiltered blue-light exposures taken by Herbig (1968) in December 1959 with the 120-in. Lick reflector, it was discovered that the bright reflection nebulosity extending northward from RY Tauri was exceptionally faint compared with plates taken in 1947 and 1957. Since the variable itself had faded to a very deep minimum in 1959, this would point to a close association between the fading of the star and that of the surface brightness of the nebula.

We must here, of course, take into account the distance between the star and this nebulosity. The distance of RY Tauri has been variously estimated at between 130 and 170 pc and at this distance, a separation of 3″ between star and nebulosity corresponds to a light-time of approximately 0.5 year. There can be, therefore, no immediate or precise response of the nebulosity to a change in brightness of the variable. From the data available, it appears probable that the decrease in surface brightness of the nebula, and possibly that of the variable also, occurred during the summer period when observations of this region are impossible due to the passage of the Sun through the constellation. Consequently, we cannot correlate these two phenomena as closely as would be desired.

CZ, DD TAURI AND IC 359

Small variations in the brightness of IC 359 have been suspected by Bohlin (1922) but there has been little confirmatory evidence of this since then. Neither of these two variables have large amplitudes, unlike RY Tauri, and unless their amplitudes in the ultraviolet (which may be a major cause of light changes in the associated nebulae) are appreciably greater than the visual amplitude, we would expect only small changes in the brightness of IC 359 due to the stars themselves.

It also appears quite possible that these two T Tauri variables are merely superimposed upon the nebula; as yet we have no accurate distance determination for either. If this is indeed so, then the $13^m.2$ star with a K type spectrum which lies 5' to the north-east of the variables may possibly be the energizing star causing the illumination of this small region of bright nebulosity.

PLATE IV. The nebulosity at RY Tauri on January 1, 1957. The scale is 7.8″/mm and north is at the top with east to the left. (Lick Observatory Photograph.)

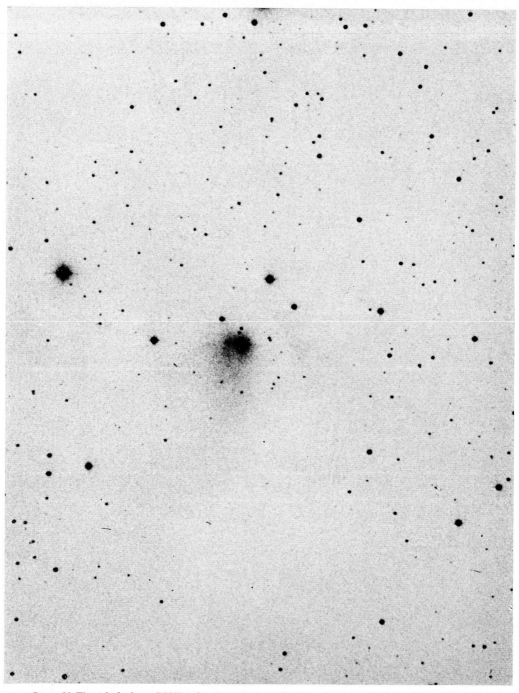

PLATE V. The nebulosity at RY Tauri on March 18, 1960. The pronounced fading of the nebulosity north and west of the variable is obvious when compared with Plate IV. (Lick Observatory Photograph.)

In this connection, however, it may be significant that DD Tauri has a very intense blue and violet continuum which almost completely obscures the underlying absorption spectrum and produces a very large colour index for this star. This particular variable is, indeed, one of the bluest of all the T Tauri stars.

DG TAURI

Barnard (1927) first discovered the nebulosity which is associated with this T Tauri variable. The shape of the nebula is very conspicuous. There is a bright portion which runs in the form of an arc to the north and east of the star. A second small patch of nebulosity about 3′ to the east contains the star Haro 6–8 and it now appears almost certain that these two nebulosities are physically connected. No definite variations in the brightness of these two nebulous regions have so far been reported.

XZ TAURI

This variable is associated with a very faint wisp of nebulosity that extends north and north-east of the star. About 23″ distant there is a very faint fifteenth magnitude companion to XZ Tauri which, on long-exposure photographs, is distinctly nebulous in outline. The light variations of this companion indicate that it, too, is a late type T Tauri variable. Changes in the brightness of this nebula have been searched for but with no conclusive results.

From the above observations it appears that, in certain cases at least, the brightness of the nebulosity and the embedded stars are related in some manner. Even in those instances where no change in the luminosity of the associated nebula has been observed, it may well be that, owing to the small amplitude of the star, no detectable change in the nebulosity is observable (or it lies at the very limit of observation). More data are required, however, before it can be positively determined whether the two effects go hand-in-hand (allowing for the light-time between star and nebula) or whether they are inversely related.

If the cause of the variations is either a purely reflection effect, or due to absorption of ultraviolet radiation from the star followed by re-emission in the visible spectrum, it is difficult to understand how they can be inversely related.

Apart from the variations in surface brightness of the nebulae, other physical interactions are, of course, possible. Among the very youngest of these stars, material will still be falling into the shock front surrounding the stellar core. We have already seen that this effect is observable in the YY Orionis variables, several of which are T Tauri variables. From their positions on the Hertzsprung–Russell diagram, these are clearly younger than the majority of these stars. The infall velocities observed by Walker (1969) are also in good agreement with the theoretical figures calculated by Larson (1972) for variables with masses between 0.25 and 1.0 $M_\odot$.

The remaining T Tauri variables which show a normal P Cygni type spectrum are older than those mentioned above and here the infall of matter has either died out or been overpowered by subsequent mass ejection.

In their case, it seems highly probable from the work of Herbig (1970) that the material being returned to the surrounding circumstellar shell has been extensively modified due to the formation of heavier elements by nuclear reactions within the star. These heavier elements, ranging in atomic number from magnesium to iron, will be preferentially in the form of solid particles, possibly as the oxides: the absorption bands of TiO, for example, are prominent in most of these M type dwarfs.

123

The Nebular Variables

In other words, we may regard these variables as chemically processing at least part of the protostellar cloud from which they originally formed and returning some of this material into the surrounding medium.

References

AMBARTSUMIAN, V. A. (1955) *Questions on Cosmogony* 4, 76, Moscow.
BARNARD, E. E. (1907) *Astrophys. J.* 25, 218.
BARNARD, E. E. (1927) *A Photographic Atlas of Selected Regions of the Milky Way*, Part II, Object No. 100, Carnegie Inst., Washington.
BOHLIN, K. (1922) *A.N.* 216, 31.
DIBAJ, E. A. (1960) *Sov. Astr. A.J.* 4, 13.
DIBAJ, E. A. (1963) *Ibid.* 7, 606.
HARO, G. (1953) *Astrophys. J.* 117, 73.
HERBIG, G. H. (1950) *Ibid.* 111, 11.
HERBIG, G. H. (1953) *A.S.P. Leaflet* No. 293.
HERBIG, G. H. (1956) *Publ. astr. Soc. Pacif.* 68, 353.
HERBIG, G. H. (1961) *Astrophys. J.* 133, 337.
HERBIG, G. H. (1962) *Adv. Astr. Astrophys.* 1, 47.
HERBIG, G. H. (1966) *Vistas in Astronomy* 8, 109, Pergamon Press, Oxford.
HERBIG, G. H. (1968) *Astrophys. J.* 152, 439.
HERBIG, G. H. (1970) *Colloque Liége*. Evolution Stellaire avant la sequence principale.
HERBIG, G. H. (1971) *Astrophys. J.* 169, 537.
HERBIG, G. H. and HARLAN, E. A. (1971) *IAU Comm.* 27, *Info. Bull.*, No. 543.
JOY, A. H. (1945) *Astrophys. J.* 102, 168.
KHACHIKYAN, E. E. and PARSAMIAN, E. S. (1965) *Astrophysics* 1, 221, Faraday Press, New York.
KUHI, L. V. (1964) *Astrophys. J.* 140, 1409.
LARSON, R. B. (1972) *Mon. Not. Roy. astr. Soc.* 157, 121.
MINKOWSKI, R. (1942) *Publ. astr. Soc. Pacif.* 54, 190.
POVEDA, A. (1965a) *Bol. Obs. Tonantzintla Tacubaya* 4, 22.
POVEDA, A. (1965b) *Ibid.* 4, 15.
WALKER, M. F. (1969) *Non-Periodic Phenomena in Variable Stars*, p. 103, Reidel, Dordrecht.

CHAPTER 19

Evolutionary characteristics

FROM a study of several T Tauri variables, Herbig (1962) has concluded that most of these stars have radii between 1 and 3 $R_\odot$, and effective temperatures in the range $3.50 < \log T_e < 3.75$. From Table V, it will be seen that these observed values and those predicted by Larson (1972a) for newly formed stars are in agreement with masses for the T Tauri variables of $\sim 1.5\ M_\odot$.

Clearly, we are here dealing with stars at the lower end of the mass range. This belief is further strengthened by the observations made by Walker (1963) of those faint T Tauri variables showing the inverse P Cygni or "YY Orionis" feature in their spectra. Not only are the positions of these stars on the Hertzsprung–Russell diagram consistent with their being among the youngest stars, but the observed infall velocities (150–400 km/sec) agree with those calculated for stars with masses between 0.25 and 1.0 $M_\odot$.

From the above observations it is clear that the remnants of the protostellar cloud are still falling into the star after the stellar core becomes visible and it is near, or just on, the main sequence. This close agreement also shows that the radii of these T Tauri stars cannot much exceed 2 $R_\odot$.

The evolution of spherical protostars having masses in the range we are now considering has been examined by Larson (1969, 1972a) and, in the main, the phases of evolution leading to the formation of a stellar core are much the same as those outlined in Chapter 7.

Non-homologous Collapse

Initially, there is a rapid condensation at the centre of the protostellar cloud followed by the formation of an opaque core which is in hydrostatic equilibrium. Dissociation of the hydrogen molecules then brings about collapse of the central region of the stellar core with the formation of a second core which has a mass of $\sim 10^{-3}\ M_\odot$. It is at this stage that the central density reaches values similar to those found in stars.

The Hertzsprung–Russell diagram for stellar cores having masses of 0.5, 1.0 and 1.5 $M_\odot$ is shown in Figs. 48–50.

The dashed portions represent the very brief phase lasting for about 1 year during which none of the radiation from the core passes through the surrounding protostellar cloud, all of it being absorbed by the infalling material. Shortly after this phase, once approximately 90 per cent of the radiated energy begins to escape as infrared radiation, the star enters upon the solid curve of its evolutionary track.

The solid dot represents the point at which half of the total mass has been accreted by the core, while the open circle indicates where virtually all of the mass has been accreted and the visual optical depth of the remnants of the cloud has diminished to about unity. This is also the point where the stellar core has evolved onto the main sequence.

The Nebular Variables

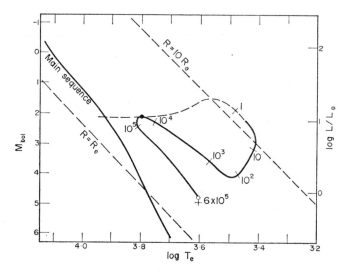

FIG. 48. The evolution of the stellar core for a protostar of 0.5 $M_\odot$. The details illustrated are those given in Fig. 31. (After Larson.)

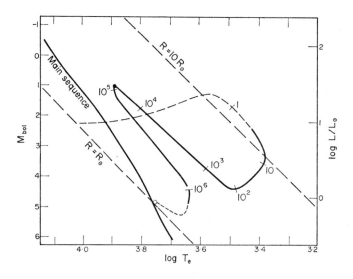

FIG. 49. The evolution of the stellar core for a protostar of 1 $M_\odot$. (After Larson.)

From Figs. 48–50 several features are immediately apparent. The evolutionary tracks for all of the stars in this mass range are almost identical during the phases when the central core is still accreting mass from the surrounding cloud. The major difference is that the surface temperature and luminosity of the core increase with an increase in mass. The radius, on the other hand, is little affected by the mass for cores of ~1.5 $M_\odot$, having a value of ~2 $R_\odot$.

We may also note that where the core has a mass of less than 1.5 $M_\odot$, the star first appears at the lower end of the"Hayashi" track, whereas with a mass of 1.5 $M_\odot$, the star is at the very end of the "Hayashi" track and the "Hayashi" phase is non-existent. This is a direct result of the stars with masses of $\lesssim 1.5\,M_\odot$ having sizeable convection zones in their outer regions.

FINAL STAGES OF THE EVOLUTION

About 8×10^4 years after the formation of the stellar core, the maximum surface temperature and luminosity are about 8300°K and $30\,L_\odot$ respectively while the core mass is $0.56\,M_\odot$. As evolution proceeds, both surface temperature and luminosity decrease as also does the opacity in the surface regions of the core. As a result, the energy outflow from the core interior becomes more dominant. With more and more of the protostellar material falling into the core, a time is reached after $\sim 10^6$ years when practically all of the protostellar material has disappeared.

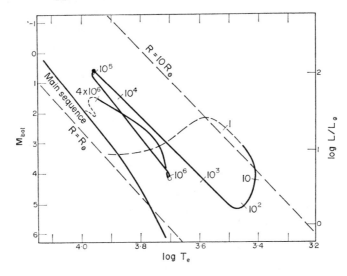

FIG. 50. The evolution of the stellar core for a protostar of $1.5\,M_\odot$. (After Larson.)

Once this stage is reached, the kinetic inflow of energy to the shock front surrounding the core reaches negligible proportions. Almost all of the energy then comes from the interior of the core. The final evolutionary tracks in Figs. 48–50, between $\sim 10^6$ and 3×10^6 years, are therefore very similar to the "Hayashi" tracks as calculated by various workers, for example Ezer and Cameron (1963, 1965), Iben (1965), and Bodenheimer (1965, 1966a, 1966b). The star itself, during this final phase, is a conventional Hayashi model in the convective phase of pre-main sequence contraction as discussed by Hayashi *et al.* (1962).

The Larson models, however, possess one distinctive feature compared with all previous models of pre-main sequence stars. Once these stars arrive on the "Hayashi" track, their radii and luminosities are of the order of $2\,R_\odot$ and $1.3\,L_\odot$. These values are considerably smaller than those previously assumed by Ezer and Cameron (1965) of $\sim 60\,R_\odot$ and $\sim 600\,L_\odot$.

Stars with Smaller Masses

Grossman (1969) has investigated the pre-main sequence and early main sequence evolution of stars having somewhat smaller masses than those considered above. Masses of

The Nebular Variables

0.06 to 0.2 $M_\odot$ have been examined with an assumed composition of $X = 0.68$, $Y = 0.29$, $Z = 0.03$ (hydrogen 68 per cent, helium 29 per cent, other elements 3 per cent).

The calculated star models are found to yield a well-defined sequence in the spectral region M4–M8, in close agreement with the observed main sequence values. The estimated contraction times to the main sequence are longer than those just considered, ranging from 3.6×10^8 years (M = 0.2 $M_\odot$) to 2.5×10^9 years (M = 0.085 $M_\odot$). For stars with the above composition, the lower mass limit on the main sequence is estimated to be 0.075 $M_\odot$.

Comparison with T Tauri Variables

Recently, the properties of the T Tauri variables have been reviewed by Kuhi (1966) and Herbig (1967). In particular, the latter has shown that the majority of these stars appear on the Hertzsprung–Russell diagram near the bottom of their "Hayashi" tracks. In addition, a large number have radii very close to 2 $R_\odot$. Both of these properties are very similar to those of the models based upon the non-homologous collapse of the protostellar cloud.

As we have also seen, evidence that these variables are still surrounded by obscuring dust present in the infalling material comes from the photometric observations of Mendoza (1966, 1968) which show a large infrared excess; indeed, there are several T Tauri variables in which the greater proportion of the total luminosity is emitted in the infrared.

In general, the infrared emission is more widely distributed in wavelength than predicted by theory. This is probably due to inhomogeneities within the infalling material or the fact that the constituent dust grains do not all evaporate at the same temperature. Taking into account the very idealized nature of the infalling cloud assumed by Larson, in the computation of the emitted spectrum of a protostar (1969), the agreement is very good.

We have already discussed the faint T Tauri variables that possess inverse P Cygni features in their spectra, interpreted as direct observational evidence for the presence of infalling material. These observations by Walker (1961, 1963, 1964, 1966, 1969) show that only a few of the fainter members exhibit this effect and that these stars also have a comparatively high ultraviolet flux. One conclusion we may draw from this is that this particular phase is of relatively short duration, occurring while the stellar core is still heavily obscured by the remnants of the infalling material.

Toma (1972) has calculated the evolutionary tracks for extreme Population I stars having masses of 0.2, 0.6, 1.0 and 2.0 $M_\odot$ in order to study the effect of deuterium burning on their pulsational stability. The initial abundance of deuterium was set to that which is generally found in meteorites.

These calculations show that vibrational instability manifests itself through the fundamental mode of the radial oscillation and, furthermore, the pulsation amplitude is multiplied by factors of between 10^8 and 10^{15} by the end of the deuterium burning phase. An attempt has been made by Toma to correlate these results with the observed properties of the T Tauri and RW Aurigae variables found in NGC 2264.

The very early stages of the collapse of a protostellar cloud have been examined by Stein, McCray and Schwartz (1972) who have demonstrated that spherically symmetric condensation, driven by thermal instabilities in a cooling interstellar medium, can produce gravitationally bound clouds so long as the magnetic field has been previously removed from the gas.

The resulting clouds have a very small core which is stationary and surrounded by a large, cool infalling envelope. The calculated parameters of the cores are: diameter $\sim 10^{-4}$ pc, temperature $100°K$ and density $\sim 10^{12}$ cm^{-3}. The surrounding envelopes are considerably larger (diameter ~ 1 pc) and cool (temperature $= 10°K$).

In many of these respects, therefore, the structures closely resemble the early stages of the protostars as envisaged by Larson (1968, 1972a).

The properties of the interstellar medium from which stars eventually evolve have been discussed by Ledoux (1971). Particular attention is paid here to the effects of gravitational and thermal instabilities which arise in this medium. The sequence of stellar evolution is followed from the evolution of collapsing clouds, through fragmentation and the formation of protostars. The effects of magnetic fields and rotation upon the evolutionary stages of such protostars are also examined.

Various other authors have recently examined the pre-main sequence evolutionary stages with reference to T Tauri variables in particular.

Bodenheimer (1972), for example, has examined the evolution of stars from just after their formation in the interstellar cloud to the point where they finally become stabilized on the main sequence. Here, nuclear reactions begin to provide the entire energy supply of the star. During the pre-main sequence stage, the star apparently shines mainly at the expense of its gravitational potential energy. Once on the main sequence, the proton–proton chain reaction begins when the central temperature is $\sim 10^7°K$ and the density ~ 100 g cm^{-3}. At somewhat higher temperatures, $\sim 2 \times 10^7°K$, the carbon–nitrogen cycle will commence.

A summary of the current observational evidence of pre-main sequence objects and their circumstellar envelopes has been given by Strom (1972), while Wenzel (1972) has discussed stellar formation with particular reference to the T Tauri and related variables in the Orion Nebula.

Effect of Rotation

So far, in discussing the evolution of the RW Aurigae, T Orionis and T Tauri variables, it has been generally assumed that these stars have formed from a spherically symmetric, non-rotating cloud. Several authors have examined the early stages of the collapse of such a cloud, e.g. Bodenheimer and Sweigart (1968), Larson (1969), Hunter (1969), Disney *et al.* (1969) and Penston (1960). While there have been various differences in the assumptions made by these workers, certain major features of the collapse appear to have been fairly well established. The cooling mechanisms prevailing at the high densities, $\sim 10^{-21}$ g cm^{-3}, are extremely efficient and maintain the temperature within the range $10–20°K$.

As a result, the collapse is isothermal, at least until the central regions become opaque to infrared radiation at a density of $\sim 10^{-13}$ g cm^{-3}. Most authors are agreed that the collapse is non-homologous although Moss (1973) has calculated various models for rapidly rotating pre-main sequence stars assuming the collapse to be homologous. Here we shall discuss these two cases separately.

HOMOLOGOUS COLLAPSE

From theoretical considerations, it is to be expected that stars which are in the contraction stages of their pre-main sequence evolution, will be rapid rotators. It is, of course, a very difficult matter to obtain direct observational evidence to confirm this prediction.

The cause of the initial rotation of a protostellar cloud has been examined by numerous workers over many decades. At present, there appear to be two possibilities.

(a) The initial rotation may occur as a direct result of galactic rotation and if this is so,

The Nebular Variables

it can readily be shown that the angular momentum will be so high that collapse into a star is impossible unless much of this angular momentum can be removed from the cloud during its collapse. The possibility that this may be brought about by the action of a magnetic field has been discussed by Mestel (1965b).

(b) Turbulent motions within the interstellar medium could lead to rotation of the protostellar clouds. This idea seems more plausible than invoking galactic rotation as the cause since there is no evidence at all that the axes of binary systems, nor stellar rotation axes, show a preferential alignment perpendicular to the galactic plane.

Unfortunately, according to this theory, we would expect the rotational velocities of protostellar clouds to be similar to their translational velocities and therefore even higher than in case (a).

Whatever the cause, it is clear that the effects of rotation of the protostellar cloud assume great importance long before stellar conditions within the cloud are attained.

The masses of the stars considered by Moss (1973) range from 0.4 to 1.0 $M_\odot$ (i.e. within the range of masses assumed for the T Tauri variables). On the assumption that the rotating protostellar cloud can dispose of a considerable proportion of its angular momentum by some means during its contraction to a stellar configuration, then this object will be fully convective and may possibly possess a large scale magnetic field. In the absence of such a field, however, the star will rotate increasingly rapidly, since convective mixing will tend to keep the rotation uniform, until the angular velocity reaches the value at which

$$\Omega^2 R_e^3 = GM \tag{1}$$

where Ω is the gravitational potential energy and R_e is the equatorial radius. Not until more angular momentum is lost can any further contraction take place. A possible mechanism, which is an idealized case, is shedding of mass from the equator, so that material rotates in Keplerian orbits in the equatorial plane of the star.

Where the star does have a large scale magnetic field, Mestel (1968) has demonstrated that the angular momentum may be effectively reduced by a stellar wind. Such magnetic fields to provide enough loss of angular momentum to maintain a constant ratio of surface gravity to centrifugal force yield relations of the form

$$R \propto M^\gamma \tag{2}$$

where γ is of the order of 100 or more. If a magnetic wind does operate, therefore, it is possible that it may be extremely efficient, certain enough to keep the angular velocity down. Even a relatively weak magnetic field could keep the angular velocity below the critical value as the star comes down the "Hayashi" track (modified for rotation).

In fact, there does seem to be some observational evidence that such a mechanism is operative in dwarf stars. As Kraft (1967) has shown, the rate of axial rotation for late dwarf stars appears to be inversely proportional to their age. The average field dwarfs rotate more slowly than those in the Hyades group (age $\sim 4 \times 10^8$ years) and these, in turn, are slower rotators than late type dwarfs in the Pleiades (age $\sim 3 \times 10^7$ years).

Following Moss (1973) we find that a sequence of stars of constant polytropic index and rotating uniformly at the critical angular velocity form an homologous sequence. It is therefore possible to calculate, by means of the normal homology relations, the changes in temperature and pressure as the masses and radii vary.

From the virial theorem for rotating stars, we have

$$2T + V + 3(\gamma - 1)U = 0 \tag{3}$$

where T is the kinetic energy, V the potential energy and U is the thermal energy.

Where $T = 0$ and $\gamma = 5/3$, we obtain the usual relations between L (the radiated luminosity) and the total energy $E = U + V$

$$L = -\frac{dE}{dt} = -\frac{1}{2}\frac{dV}{dt} = \frac{dU}{dt}. \tag{4}$$

With constant mass and assuming homologous contraction, we have

$$L = \frac{3}{7}\frac{GM^2}{R}J \tag{5}$$

where

$$J = -\frac{1}{R}\frac{dR}{dt}. \tag{6}$$

If $T \neq 0$, then

$$E = T + V + U = -[T + (3\gamma - 4)U] \tag{7}$$

which, using equation (3) and $\gamma = 5/3$, reduces to

$$L = \frac{d}{dt}(T + U) = -\frac{1}{2}\frac{dV}{dt}. \tag{8}$$

Although it is not easy to evaluate V for the general rotating configuration, U can be evaluated, the final result being

$$L = \frac{JGM^2}{R_e}\left(\frac{3q_n}{2} + \frac{k^2}{2}\right) \tag{9}$$

where M is constant.

In the case where the mass is not constant, it is necessary to modify equation (9) by subtracting the energy flux necessary to separate the mass lost from the star. Idealizing the situation, we may consider two cases.

(a) During contraction, the star sheds mass from the equator, leaving a ring of material moving in Keplerian orbits and

(b) the mass is ejected from the star to infinity.

In either case, we may write the modified equation as

$$L = \frac{JGM^2}{R_e}\left(\frac{3q_n}{2} + \frac{k^2}{2}\right)\left(1 - 2\frac{d\log M}{d\log R_e}\right) + j\frac{d\log M}{d\log R_e}. \tag{10}$$

In case (a) referred to above, $j = 0$, whereas in case (b), $j = \frac{1}{2}$. The values for q_n and k^2 in the above equations may be readily obtained by numerical integration.

Both non-rotating and rotating models have been calculated by Moss using a composition of $X = 0.70$, $Y = 0.28$, $Z = 0.02$ and masses of 0.4, 0.5 and 1.0 $M_\odot$.

On the Hertzsprung–Russell diagram, it is found that the effect of rotation is to shift the models towards lower effective temperatures and luminosities at the same volume (Fig. 16). From the results obtained, it also appears that one further effect is to increase the contraction time quite appreciably compared with non-rotation models.

The Nebular Variables

In this case, the density distribution becomes sharply peaked at the centre and the density law has the form

$$\rho \propto r^{-2}. \tag{11}$$

This characteristic of the collapsing cloud assumes importance in the later evolutionary stages of the protostar. Larson (1972b) has examined variations from spherical symmetry of the protostar cloud for both non-rotating and rotating models. In the former case, it has been demonstrated by Lynden-Bell (1964) and Mestel (1965b) that the eccentricity of a uniform, pressure-free spheroid collapsing under gravity increases gradually during collapse. Calculations by Lin, Mestel and Shu (1965) show that a prolate spheroid collapses eventually to a line while an oblate spheroid collapses rapidly to a disc. We must also take into account the possibility of fragmentation during collapse which also constitutes a deviation from spherical symmetry.

From the calculations made by Larson (1972b) it appears that where there is no rotation, a gravitationally collapsing isothermal cloud does not become unstable to fragmentation as the collapse proceeds and the pressure forces remain sufficient to prevent any large deviation from spherical symmetry. Only in the case where the initial temperature is as low as 7.5°K does fragmentation take place with the formation of two density maxima.

Similarly, fragmentation can occur during the early stages of the collapse of a massive, low-density protostellar cloud where, as shown by Hunter (1969) and Disney *et al.* (1969), the temperature falls rapidly with an increase in the density.

There is also no *a priori* reason why, if two condensations are formed in a non-rotating cloud, they should not coalesce into a single condensation.

For a rotating protostellar cloud, however, the situation is quite different. In the models computed by Larson, a mass of $1\,M_\odot$ and a temperature of 10°K have been assumed, the initial density distribution is uniform and a uniform (solid body) rotation has been adopted.

One important result of these calculations is that there is no collapse unless the initial angular velocity ω_0 is smaller than a critical velocity ω_c that is dependent upon the radius of the cloud. The values of ω_c for different values of the radius R, as found by Larson, are given in Table XV.

TABLE XV. CRITICAL ANGULAR VELOCITIES

R (cm)	R/R_m	ω_c (sec^{-1})	ω_c/ω_m
1.0×10^{17}	0.59	3.2×10^{-13}	0.9
1.4×10^{17}	0.82	1.1×10^{-13}	0.5
1.63×10^{17}	0.96	3.3×10^{-14}	0.2

In the above table, R is expressed in terms of $R_m = 1.7 \times 10^{17}$ cm, this being taken as the maximum radius for which collapse can take place in the absence of rotation, and ω_c is expressed in terms of $\omega_m = (4\pi G \rho_0/3)^{\frac{1}{2}}$ which is the value for ω_0 where gravity and centrifugal force just balance in the equatorial plane. This is, of course, analogous to the situation we found earlier for the homologous collapse of a protostar.

The calculations show that, in general, the deviations from spherical symmetry here increase steadily during the collapse and the central regions eventually fragment into two or more condensations in orbit about each other. It is quite possible, therefore, that rotation may be the major cause of fragmentation during the early stages of stellar evolution.

Prior to fragmentation into condensations, it appears that a ring will be formed in the centre of the cloud. If the collapse is non-homologous, the formation of such a ring may be a general result of the collapse of a rapidly rotating cloud. It seems likely that this ring will be extremely unstable and may never really form at all as suggested from the work of Lyttleton (1953) and Arny (1967).

One important outcome of this work is that the formation of binary or multiple systems in this manner largely overcomes the difficulty of the dissipation of the angular momentum of such a cloud since this will go into the orbital angular momentum of the stars themselves. It also provides a plausible explanation of the high number of binary and multiple systems which have been observed (cf. Heintz, 1969).

However, can we explain the numerous single stars on this basis? The possibility put forward by Larson is that in those cases where several stars are formed from a single cloud, a process of ejection of single stars takes place until only a stable binary remains. The *n*-body problem has been studied and the calculations carried out by Agekyan and Anosova (1968), these illustrating, quite clearly, the instability of such systems.

It appears natural that the least massive stars would be preferentially ejected, leaving the more massive behind in the resulting binaries. There is some observational evidence in support of this idea from the work of Blaauw (1961) who has shown that the O-type stars which are young, massive stars, are nearly always found as binaries. Except for the run-away O-type stars, single stars of this spectral type are extremely rare and even in the cases of the run-away stars, it is believed that these originated in binary systems and were ejected by explosive mass loss from one of the components.

It is also significant that we find a number of binaries in which at least one of the components is a nebular variable. Some typical binaries are given in Table XVI.

TABLE XVI. NEBULAR VARIABLES IN BINARY SYSTEMS

Star	Spectrum	Magnitude	Companion	Spectrum	Magnitude	Sep
DD Tau	dK6e	14.5–15.5	CZ Tau	dM2e	15.8–17.3	31″
DH Tau	dM0e	14.2–15.1	DI Tau	dM0e	14.3–15.6	16″
UY Aur	dG5:e	11.6–14.0	—	e	12.1	0.8″
S CrA	e	10.9–12.6	—	—	13.5	1″

The Angular Momentum Problem

A somewhat different approach to the problem of disposing of the excess angular momentum in a collapsing protostellar cloud, and the associated one of the contraction of the magnetic field, has been made by Prentice and ter Haar (1971). The idea that star formation occurs in collapsing dust clouds rather than gas clouds was put forward by Reddish and Wickramasinghe (1968, 1969) although these authors did not take the angular momentum problem into account.

Mechanisms for disposing of the required angular momentum in a gas cloud are very difficult to find. Ter Haar (1949) and Hoyle (1960) have shown that hydrodynamic forces are insufficiently strong. According to Ebert *et al.* (1960), even magnetic forces appear to be inadequate. The possibility that a neutral gas cloud may collapse across the magnetic field lines due to a recombination of ions and electrons has been put forward by Mestel and Spitzer (1956) and Spitzer (1968). The adequacy of this process also appears unlikely when one takes account of the ionization produced by cosmic rays as shown by Pottasch (1968), Pikel'ner (1968) and Prentice and ter Haar (1969). Indeed, Prentice (1970) has demonstrated that the cosmic ray intensity at the centre of a gas cloud of 1 $M_\odot$ does not show any appreciable diminution until the radius of the cloud has collapsed to ~ 100 a.u.

In a collapsing dust cloud, however, the problems of angular momentum and magnetic flux are not as serious. Reddish and Wickramasinghe (1968, 1969) have shown that in a cloud with a hydrogen atom density of $\sim 10^4$ g cm^{-3} it is possible for solid hydrogen grains to form and grow to a radius of $\sim 10^{-4}$ cm. The temperatures within the cloud will be of the order of 3°K and consequently no helium will condense. Assuming the normal cosmic abundance ratio of hydrogen/helium, the maximum mass of the cloud that will condense will be about 70 per cent. Once the cloud has condensed in this way, there will be no coupling to the magnetic field and, as a result, the magnetic flux problem disappears.

Prentice and ter Haar (1971) have considered a protostellar cloud of $\sim 10^4$ $M_\odot$ which is rotating with an angular velocity of 3×10^{-16} sec^{-1} and collapsing until its density reaches approximately 10^4 hydrogen atoms per cm^3. Little angular momentum can be lost by the cloud during this stage at its collapse, but it can be demonstrated that, as yet, it will not have become rotationally unstable. From the analysis carried out by Hunter (1962), however, it does appear probable that fragmentation of the cloud will have occurred.

If a fragment of ~ 1 $M_\odot$ is now considered, we find that its mass is not large enough for it to be able to condense against the gas and magnetic pressure if it is composed solely of gas. The solid hydrogen grains, however, will fall towards the centre of the fragment and lose angular momentum by interaction with the neutral and ionized gas, the latter being mainly helium.

From their calculations, Prentice and ter Haar find that during the first stage of the collapse of a grain cloud it is possible for it to lose a large fraction of its angular momentum and that for the first 10^7 years or so the collapse will proceed such that the grains appear to fall down the spokes of a wheel that rotates with the same constant angular velocity as that of the original cloud in which the grains were formed.

It is, of course, essential that the transfer of the angular momentum to the outside must be sufficiently rapid. Since the speed with which the angular momentum is transferred will be essentially the Alfven velocity, this may be calculated and compared with the radius of the 1 $M_\odot$ fragment. The time scale involved is thus found to be $\sim 10^3$ to 10^4 years which is considerably shorter than the time scale of the initial collapse phase.

Mass Loss in Contracting Pre-Main Sequence Stars

An attempt to simulate the T Tauri phase of pre-main sequence evolution by including a simple scheme for mass loss has been made by Kuhi and Forbes (1970). These authors have calculated four stellar models for intercomparison; two with an initial mass of 1.5 $M_\odot$ and final mass 1.2146 $M_\odot$ with rates of mass loss estimated by different prescriptions, and two with constant masses at the above two values.

The final zero age main sequence position of the cases with mass loss corresponds exactly to the final mass but, more surprisingly, the time scale for contraction to the main sequence also virtually corresponds to the final mass. Clearly, the higher mass loss during the early

"Hayashi" phase has little effect. An idealized treatment of the mass ejection suggests that for observationally reasonable rates of mass loss, the observed effective temperatures may attain low enough values to be well within the Hayashi forbidden region on the right of the Hertzsprung–Russell diagram which is set by the fully convective condition. Some of the features observed in the Hertzsprung–Russell diagram for the stars in NGC 2262 seem explainable on the basis of mass loss during the pre-main sequence phase.

Ezer (1971) has provided a semi-empirical formulation of the rate of stellar mass loss by the stellar wind. Evolutionary studies of the T Tauri phase are presented for a variety of rates of mass loss. It is found that different mass loss rates produce only very small changes in the positions of equal evolutionary time lines in the Hertzsprung–Russell diagram. From this work it is concluded that the spread of points in the Hertzsprung–Russell diagrams of young clusters results from the spread in their formation times. This is consistent with the initiation of star formation by the violent hydrodynamic compression of a typical interstellar cloud.

Nobile and Secco (1970), on the other hand, have studied a set of evolutionary models with different masses which happen to be unstable against rotation in the phases of pre-main sequence contraction, choosing the angular momentum as the free parameter. The models are taken as having an initial mass of $2\,M_\odot$ with an extreme Population I composition ($X = 0.602$, $Y = 0.354$, $Z = 0.044$). The distribution, by mass, of the isotopes He^3, C^{12}, N^{14} and O^{16} is chosen to be $X_3 = 0.00$, $X_{12} = 0.00619$, $X_{14} = 0.00204$ and $X_{16} = 0.01847$ respectively. The evolutionary track has been studied from near the top of the corresponding "Hayashi" line where the radius is $\sim 50\,R_\odot$. The equatorial velocity is assumed to be 94 km/sec (corresponding to an angular moment $I = 1.92 \times 10^{52}$ cgs). The requirement of a balance between centrifugal and gravitational forces at the limiting outer surface leads inevitably to mass loss during the contraction down to lower radii.

References

AGEKYAN, T. A. and ANOSOVA, ZH. P. (1968) *Astrophysics* **4**, 11.
ARNY, T. (1967) *Ann. Astrophys.* **30**, 1.
BLAAUW, A. (1961) *Bull. astr. Inst. Netherlands* **15**, 265.
BODENHEIMER, P. (1965) *Astrophys. J.* **142**, 451.
BODENHEIMER, P. (1966a) *Ibid.* **144**, 103.
BODENHEIMER, P. (1966b) *Ibid.* **144**, 709.
BODENHEIMER, P. and SWEIGART, A. (1968) *Ibid.* **152**, 515.
BODENHEIMER, P. (1972) *Rep. Prog. Phys.* **35**, 1.
DISNEY, M. J., McNALLY, D. and WRIGHT, A. E. (1969) *Mon. Not. Roy. astr. Soc.* **146**, 123.
EBERT, R., HOERNER, S. VON and TEMESVARY, S. (1960) *Die Entstehung von Sternen durch Kondensation diffuser Materie*, p. 311, Springer, Heidelberg.
EZER, D. and CAMERON, A. G. W. (1963) *Icarus*, **1**, 422.
EZER, D. and CAMERON, A. G. W. (1965) *Can. J. Phys.* **43**, 1487.
EZER, D. (1971) *Astrophys. and Space Sci.* **10**, 52.
GROSSMAN, A. S. (1969) *Am. Astron. Soc.* **13**, 291.
HAAR, D. TER (1949) *Astrophys. J.* **110**, 321.
HAYASHI, C., HOSHI, R. and SUGIMOTO, D. (1962) *Prog. Theor. Phys. Suppl.* No. 22.
HEINTZ, W. D. (1969) *J. Roy. astr. Soc., Canada* **63**, 275.
HERBIG, G. H. (1962) *Adv. astr. Astrophys.* **1**, 47.
HERBIG, G. H. (1967) *Scient. Am.* **217**, 30.
HOYLE, F. (1960) *Q. J. Roy. astr. Soc.* **1**, 28.
HUNTER, C. (1962) *Astrophys. J.* **136**, 594.
HUNTER, J. H. (1969) *Mon. Not. Roy. astr. Soc.* **142**, 473.
IBEN, I. (1965) *Astrophys. J.* **141**, 993.
KRAFT, R. B. (1967) *Ibid.* **135**, 748.

The Nebular Variables

KUHI, L. V. (1966) *J. Roy. astr. Soc., Canada* **60**, 1.
KUHI, L. V. and FORBES, J. E. (1970) *Astrophys. J.* **159**, 871.
LARSON, R. B. (1969) *Mon. Not. Roy. astr. Soc.* **145**, 271.
LARSON, R. B. (1972a) *Ibid.* **157**, 121.
LARSON, R. B. (1972b) *Ibid.* **156**, 437.
LEDOUX, P. (1971) *Structure and Evolution of the Galaxy*, p. 208, Reidel, Dordrecht.
LIN, C. C., MESTEL, L. and SHU, F. H. (1965) *Astrophys. J.* **142**, 1431.
LYNDEN-BELL, D. (1964) *Ibid.* **139**, 1195.
LYTTLETON, R. A. (1953) *The Stability of Rotating Liquid Masses*, Cambridge Univ. Press.
MENDOZA, E. E. (1966) *Astrophys. J.* **143**, 1010.
MENDOZA, E. E. (1968) *Ibid.* **151**, 977.
MESTEL, L. and SPITZER, L. (1956) *Mon. Not. Roy. astr. Soc.* **116**, 503.
MESTEL, L. (1965a) *Q. J. Roy. astr. Soc.* **6**, 161.
MESTEL, L. (1965b) *Ibid.* **6**, 265.
MESTEL, L. (1968) *Mon. Not. Roy. astr. Soc.* **138**, 359.
MOSS, D. L. (1973) *Ibid.* **161**, 225.
NOBILE, L. and SECCO, L. (1970) *Mem. Soc. Roy. Sci. Liége* **19**, 207.
PENSTON, M. V. (1969) *Mon. Not. Roy. astr. Soc.* **145**, 457.
PIKEL'NER, S. B. (1968) *Sov. astr. A. J.* **11**, 737.
POTTASCH, S. R. (1968) *Bull. astr. Inst. Netherlands* **19**, 469.
PRENTICE, A. J. R. and HAAR, D. TER (1969) *Acta. Phys. Acad. Sci. Hungary* **27**, 231.
PRENTICE, A. J. R. (1970) *Thesis*, Oxford.
PRENTICE, A. J. R. and HAAR, D. TER (1971) *Mon. Not. Roy. astr. Soc.* **151**, 177.
REDDISH, V. C. and WICKRAMASINGHE, N. C. (1968) *Nature* **218**, 661.
REDDISH, V. C. and WICKRAMASINGHE, N. C. (1969) *Mon. Not. Roy. astr. Soc.* **143**, 189.
SPITZER, L. (1968) *Diffuse Matter in Space*, Interscience, New York.
STEIN, R. F., McCRAY, R., SCHWARTZ, J. (1972) *Astrophys. J. Lett.* **177**, L125.
STROM, S. E. (1972) *Bull. Am. astr. Soc.* 4, 324.
TOMA, E. (1972) *Astron. and Astrophys.* **19**, 76.
WALKER, M. F. (1961) *Comptes Rendus* **253**, 383.
WALKER, M. F. (1963) *Astrophys. J.* **68**, 298.
WALKER, M. F. (1964) *Roy. Obs. Bull.* No. 82, 69.
WALKER, M. F. (1966) *Stellar Evolution*, p. 405, Plenum Press, New York.
WALKER, M. F. (1969) *Non-Periodic Phenomena in Variable Stars*, p. 103, Reidel, Dordrecht.
WENZEL, W. (1972) *Jena Rev.* **17**, 319.

CHAPTER 20

Evidence of protoplanetary systems

IN THE formation of a planetary system, one prerequisite is the presence of a protoplanetary disc of material consisting of both gas and solid matter. The possibility that some of the nebulae associated with certain nebular variables have evolved in such a manner that they are concentrated as a disc along the equatorial plane of the star has been put forward by various authors, including Poveda (1965) and Herbig (1968, 1971).

This idea has the advantage that it satisfactorily explains the peculiar biconical nebulae associated with stars such as LkHα-208 whose geometry is very suggestive of a disc of obscuring material lying in the equatorial plane of the star. This will efficiently absorb the radiation from this stellar zone but allow the light from the polar regions to be scattered by the surrounding medium.

Many of these variables also show the normal P Cygni type of absorption spectrum with the lines of hydrogen and Ca II being shifted towards the violet borders of the emission lines. The most likely explanation of this is ejection of matter into the surrounding region and, as these objects are rapidly rotating stars, we may expect this ejection to take place preferentially in the equatorial plane.

Formation of a Protoplanetary Disc by Accretion

The fact that, by the time the nebular variables approach the main sequence they are still surrounded by the remnants of the protostellar cloud from which they condensed, implies that a protoplanetary disc may come into being by an accretion process analogous to that proposed by Lyttleton (1961) and extended by Aust and Woolfson (1973) for the formation of the solar system. This supposes that if a star of $1\,M_\odot$ moves through an interstellar cloud, it will form a disc of material by an accretion process. Two mechanisms are relevant here.

(a) If the star is at rest with respect to the cloud which has a thermal velocity at large distances $= c$, then according to the case of spherically symmetric accretion due to Bondi (1952), all material within a radius R_c will be captured by the star, where

$$R_c = \frac{2GM_\odot}{c^2}. \tag{1}$$

(b) If, on the other hand, the star is travelling with a velocity V with respect to the cloud (and the various temperature effects are ignored), then line accretion (Bondi and Hoyle, 1944) will take place and material will be accreted with a radius R_v where

$$R_v = \frac{2GM_\odot}{V^2}. \tag{2}$$

The Nebular Variables

Equation (2) has been modified by Bondi (1952) to take the temperature effects into account and the capture radius R is then given by

$$R = \frac{2GM_\odot}{V^2 + c^2}.$$ (3)

Clearly, in the case of a nebular variable situated within a rotating cloud, equation (1) will apply. This enables us to ignore the tidal distortions of the cloud produced by the gravitational field of the star as in the case of the Sun approaching and passing through such a cloud discussed by Aust and Woolfson. The end result, however, will be a spherical circumstellar envelope such as is found for many of these stars.

For disc formation we must consider the process of line accretion where the star is moving through the cloud. From the theoretical results obtained by Larson (1972) on the final fragmentation stage during the collapse of a rotating protostellar cloud it seems likely that a number of accreting cores will be formed moving through an extended cloud of uncondensed material. Unless an O-type star is formed (when most of the remaining material will be blown away by radiation pressure) accretion will continue until all of the remaining material has become exhausted.

On Lyttleton's theory, the gravitation field of the star deflects the streaming material into hyperbolic paths. At some region behind the moving star, collisions between these streams will result in the destruction of their angular momentum. If, now, the remaining radial component of velocity is not sufficient to allow escape of the material, it will infall towards the star. The initial angular momentum of the material, however, will prevent direct infall and instead the gas and dust will go into orbit forming a disc in the invariant plane which is uniquely defined by the angular momentum vector of the protostellar cloud.

The characteristics of the cloud considered by Lyttleton are similar to those assumed by Larson (1969, 1972) for the formation of a protostar, namely all of the hydrogen is in the molecular form, dust comprises ~ 1 per cent by mass of the cloud and the material is in thermal equilibrium with the background galactic radiation at $3.18°K$.

The validity of the assumption that all of the hydrogen is present as molecules has, however, been questioned by Stecher and Williams (1967). Although the probability of atomic hydrogen forming molecules on the surface of a dust grain is high where the temperature of the grain is $\sim 20°K$, at least 10 per cent of the excitations of molecular hydrogen brought about by photons having wavelengths greater than 91.2 nm bring about dissociation. Such photons are present in normal H I regions. This being the case, the abundance of molecular hydrogen in protostellar clouds is probably only a small fraction of the atomic abundance.

Efficiency of the Accretion Mechanism

Since the theory of accretion mechanisms was put forward by Lyttleton (1961) and Bondi and Hoyle (1944), two other kinds of accretion process have been examined in detail.

Danby and Camm (1957) have studied the case where molecular collisions within the accreted material may be neglected. In this case, the mean free path of the molecules is relatively large compared with the capture radius. Here we find a Maxwellian velocity distribution in the material of the cloud. This will result in the majority of molecules following orbits about the star, but in general, these do not intersect the accretion line behind the moving star. Consequently, the efficiency of the accretion mechanism is quite an inefficient process under these conditions.

However, in the case of the dense dust clouds we are at present considering, it appears far more likely that collisions among the hydrogen molecules and dust grains will be important. In this particular case, the flow of the material around the moving star may be treated according to fluid dynamics as examined by Hunt (1971) for the case of a galaxy moving through an intergalactic plasma.

Recent determinations of the total collision cross-sections for H–H and H_2–H_2 collisions have been made by Massey (1971) which show that these are ~ 100 times larger than those derived earlier by Danby and Camm from the classical radius of the hydrogen atom. These larger cross-sections hold for temperatures below $100°$K (i.e. in the temperature range we are considering here). Under these conditions, the mean free path of a molecule is far smaller than the capture radius, thereby suggesting that accretion by this process will be sufficiently efficient to produce a protoplanetary disc provided that (a) the relative velocity of star and cloud is sufficiently small and (b) thermal forces do not bring about extensive loss of material by increasing the orbital velocity beyond that required for escape from the star.

Aust and Woolfson have examined the accretion process over a wide range of radii, densities and temperatures for the surrounding cloud assuming sphericity and non-distortion by the accreting body. Over quite a range of densities and radii, assuming $20°$K as the temperature of the cloud, they find that material having a mass and angular momentum similar to that of the solar planetary system can be captured on a time scale of $\sim 1.5 \times 10^6$ years assuming an initial density of 10^{-19} g cm^{-3}. This is of the same order of time as the age of most of the nebular variables.

Formation of a Planetary Disc by Capture

The capture theory for planetary formation was proposed by Woolfson (1964) and later expanded by Dormand and Woolfson (1971). According to this hypothesis, the material forming the protoplanetary disc is captured by the star from a protostar in the pre-main sequence stages of its evolution, surface temperature $\sim 50°$K and density $\sim 10^{-8}$ g cm^{-3}.

If we wish to apply this theory to the formation of protoplanetary systems around nebular variables, we must assume that stellar formation is going on continuously within these clouds (this is generally in good agreement with observation). A close encounter between the early stage protostar and the nebular variable will result in the latter drawing out a tongue of material from the former by tidal action.

As the young protostar recedes, protoplanetary condensations will be formed within the tidally raised filament, these being drawn into orbit around the star. As we have already seen, the final result of the non-homologous collapse of a rapidly rotating protostellar cloud is likely to be a number of accreting cores moving within the centre of the cloud. The chances of such an encounter are therefore quite high in these regions. A difficulty associated with this theory is that the protoplanets are formed directly during the capture mechanism and not by later, subsidiary processes as in the accretion theory. In the case of LkHα-208, there is no observational evidence at all for the presence of discontinuities within the ring of obscuring matter around the central star.

Arny and Weissmann (1973) have studied the collapse and early evolution of a cluster of protostars using an *n*-body integration. On this basis, each protostar is assigned an initial radius which decreases rapidly with time. As the cluster collapses, a random transverse velocity grows rapidly giving rise to orbital mixing of the protostars which prevents the fragments from reaching the centre of the cloud simultaneously.

It can be shown that approximately 50 per cent of all the evolving protostars will suffer very close encounters or even direct physical collisions during the initial collapse of the

The Nebular Variables

cloud provided that the radii of the protostars do not shrink more than four times as fast as the cluster itself.

One particularly important outcome of this theory is that angular momentum is lost to the surrounding medium and this occurs somewhat more rapidly than has previously been supposed. During the early stages of its evolution, therefore, a protostar can gain quite large additional amounts of mass via inelastic collisions with other protostars in the cloud.

It appears quite probable, therefore, that much of this additional mass may end up in the form of a protoplanetary disc, possibly by the mechanism suggested earlier.

Formation of a Protoplanetary Disc by Mass Ejection

Since several of these nebular variables have been shown to be ejecting material into the surrounding medium (from their violet-displaced absorption components) it is feasible that such material may eventually form into a disc-shaped structure lying in the equatorial plane of the star. We have already seen that the majority of these pre-main sequence stars appear to have relatively high rates of axial rotation.

As the central star contracts, more and more of its angular momentum will be transferred into the rotating disc. At this point, we encounter a serious difficulty. At some stage, this disc will become separated from the star and some means must be found for the rotational momentum to be transferred across the ensuing gap. It is now considered that the presence of a magnetic field surrounding the star will not only allow rotational momentum to be carried across the gap but will also have the additional effect of reducing the axial rotation of the star while at the same time forcing much of the gaseous material further away.

This process is known to work perfectly well with gases but it will not do so with solids. As a consequence of this theory, therefore, the solid grains in the surrounding ring will remain in the zone nearest the star while the lighter gases are expelled to much greater distances. It is perhaps significant that this is exactly the situation we find at present in the solar system with the inner, terrestrial planets containing a high proportion of silicate rock and iron; the more distant planets being composed predominantly of hydrogen, ammonia, methane and carbon dioxide together with the inert gases.

Which of the two processes, accretion or ejection, could predominate in the case of the T Tauri variables will depend upon the mass of dust remaining in the protostellar cloud by the time the central core approaches the main sequence and is then, essentially, a star.

References

ARNY, T. and WEISSMANN, P. (1973) *Astron. J.* **78**, 309.
AUST, C. and WOOLFSON, M. M. (1973) *Mon. Not. Roy. astr. Soc.* **161**, 7.
BONDI, H. and HOYLE, F. (1944) *Ibid.* **104**, 273.
BONDI, H. (1952) *Ibid.* **112**, 195.
DANBY, J. M. A. and CAMM, G. L. (1957) *Ibid.* **117**, 50.
DORMAND, J. R. and WOOLFSON, M. M. (1971) *Ibid.* **151**, 307.
HERBIG, G. H. (1968) *Astrophys. J.* **152**, 439.
HERBIG, G. H. (1971) *Ibid.* **169**, 537.
HUNT, R. (1971) *Mon. Not. Roy. astr. Soc.* **154**, 141.
LARSON, R. B. (1969) *Mon. Not. Roy. astr. Soc.* **145**, 271.
LARSON, R. B. (1972) *Mon. Not. Roy. astr. Soc.* **157**, 121.
LARSON, R. B. (1972) *Ibid.* **156**, 437.
LYTTLETON, R. A. (1961) *Ibid.* **122**, 399.
MASSEY, H. S. W. (1971) *Electronic and Ionic Impact Phenomena*, Vol. III, Clarendon Press, Oxford.
POVEDA, A. (1965) *Bol. Obs. Tonantzintla Tacubaya* **4**, 15.
STECHER, T. P. and WILLIAMS, D. A. (1967) *Astrophys. J. Lett.* **149**, L29.
WOOLFSON, M. M. (1964) *Proc. Roy. Soc. A.* **282**, 485.

PART IV

Peculiar Nebular Variables

CHAPTER 21

R Aquarii

THE LONG period variable R Aquarii has an extreme visual range of $6^m.3$–$11^m.8$ and a mean period of 386.86 days. There are, however, several peculiarities associated with both its light and spectroscopic characteristics which indicated, many years ago, that it may not be a typical member of this class.

Unfortunately, R Aquarii is a zodiacal variable and consequently it is unobservable for 2 or 3 months of the year when the Sun passes through this particular constellation. Otherwise, however, the light curve (Fig. 51) is complete for several decades, due mainly to the efforts of the American Association of Variable Star Observers (1973).

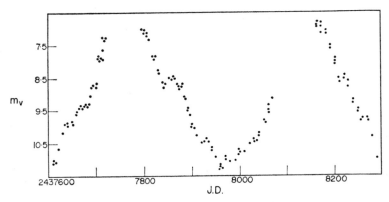

FIG. 51. Light curve of R Aquarii.

Figure 51 shows that this star exhibits most of the features of stars of the long period variable class with the maxima, in general, occurring close to the mean period. The O–C diagram (Fig. 52) shows this consistency quite clearly. The heights of the individual maxima vary quite appreciably, often by between $1^m.0$ and $2^m.0$, whereas the minima are more or less constant in brightness around $11^m.5$. Very pronounced distortions appear on both the ascending and descending branches of the light curve, usually being located midway between maximum and minimum. On a few occasions, as during the maximum of JD 2436610, the hump appears just before maximum brightness.

In all of these respects, R Aquarii behaves very like the majority of the Mira variables which also show similar irregularities in their light variations. An examination of the photographic amplitude of this star, however, shows that it varies between $6^m.7$ and $11^m.6$. This fact, in itself, may not appear to be so remarkable until we recall that all of the long period variables are red stars of spectral types M, S, N, R or C (or have intermediate spectra of types SC and CS). The spectrum of R Aquarii is predominantly type M7e with broad,

The Nebular Variables

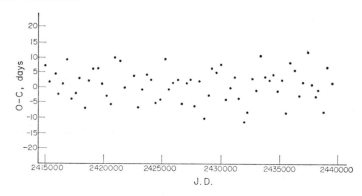

Fig. 52. Observed minus calculated dates of maxima of R Aquarii to the turn of the century.

fluted absorption bands of TiO and emission lines of the Balmer series when around maximum light.

Red stars such as this usually possess quite large colour indices of the order of 1 or 2 magnitudes which manifest themselves both in the magnitude at maximum and minimum as may be seen from a comparison of the visual and photographic ranges of the typical long period variables given in Table XVII.

Quite clearly, R Aquarii differs from these variables in that the visual and photographic light ranges are almost identical, suggesting a colour index close to zero for this star. From

TABLE XVII. VISUAL AND PHOTOGRAPHIC RANGES OF LONG PERIOD VARIABLES

Star	Visual range		Photographic range		Spectrum
	Maximum	Minimum	Maximum	Minimum	
R Aqr	6.3	11.8	6.7	11.6	M7e*
S Aqr	7.2	13.5	9.3	15.6	M4e
U Ara	7.6	14.0	9.2	15.4	M4e
R Car	4.0	9.9	5.6	11.1	M5e
S Car	5.4	10.0	6.9	11.0	M0e
R Cen	5.1	11.8	7.7	12.3	M4e
U Cen	7.0	13.6	9.3	15.3	M3e
W Cen	6.6	13.7	8.6	15.5	M3e
RS Cen	7.8	14.2	9.2	15.3	M3e
R Cha	7.1	14.1	8.5	15.5	M7e
R Gru	7.2	14.6	8.7	16.9	M5e
S Gru	6.2	14.4	7.3	15.8	M6e
R Hor	4.6	13.9	6.3	15.0	M7e
S Hor	8.1	13.6	9.7	15.2	M6e
T Hor	7.1	13.8	9.2	15.0	M4e
R Nor	6.6	13.8	8.5	14.3	M4e
R Phe	7.5	14.9	9.2	15.3	M3e
S Pic	7.8	14.1	9.3	16.2	M7e
R Ret	6.6	13.7	8.3	14.6	M4e
T Tuc	7.6	14.0	9.3	15.5	M2e
W Vel	7.9	13.8	9.5	15.1	M7e

*We shall see later that the spectrum is considerably more complex than this.

the few photographic observations which are available for R Aquarii, it would seem that the visual and photographic light curves are almost superimposable (Fig. 53).

Clearly, this small colour index of R Aquarii cannot be explained simply on the basis of observational error and we must seek its cause elsewhere. Fortunately, the necessary clues are provided by high dispersion spectrograms of this star which is sufficiently bright at all phases for these to be readily obtained.

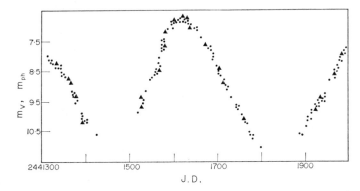

FIG. 53. Visual and photographic light curves of R Aquarii. The solid dots represent visual estimates; the solid triangles represent photographic determinations.

The Spectrum of R Aquarii

At high dispersion, the spectrum of this variable is found to be extremely complex with various absorption bands due to TiO which vary in width and intensity with the fluctuations in brightness. There are also emission lines of the Balmer series which appear on the rise to maximum, attain their greatest intensity shortly after maximum light, and then fade as the brightness declines. These are, of course, typical features of stars having M7e-type spectra.

In addition, however, there is a second spectrum superimposed upon this consisting of absorption lines due to ionized atoms, particularly those of He II, O II and N II. This is the type of spectrum found in Type O stars with high surface temperatures of the order of 30,000 to 35,000°K. Nor is this all, for a small number of emission lines are also present which possess absorption components on their violet borders. Such Doppler-shifted absorption lines are characteristic of the small group of P Cygni variables.

P Cygni Variables

We may here examine the P Cygni stars in some detail since they have some bearing upon the picture we have of R Aquarii and the reason for its inclusion among the nebular variables. The prototype star, P Cygni, was first recorded in 1600 as a third magnitude nova by Blaeu. In 1606 it began a slow decline until two decades later it was invisible to the naked eye. In 1654, it commenced to brighten once more to just above $6^m.0$ and, after a series of minor fluctuations, it finally attained $5^m.2$ where it has remained ever since with only very small changes in brightness still taking place.

Some observers, for example Nikonov (1936, 1937), have reported irregular variations with an amplitude of $\sim 0^m.2$ and more recently, Magalashvili and Kharadze (1967a, 1967b) have carried out two and three colour photometric observations of this star covering the period from 1951 to 1960. From their results, they concluded that P Cygni is a W Ursae

145

The Nebular Variables

Majoris type eclipsing binary with a period $P = 0.500656$ days and amplitudes of $0^m.10$ and $0^m.08$ for the primary and secondary minima respectively.

Further observations by Alexander and Wallerstein (1967) have failed to substantiate this conclusion. There is no doubt that P Cygni is both large and bright and from its position on the Hertzsprung–Russell diagram, the radius must be at least $100\ R_\odot$. Theoretical calculations show that for a primary with so large a radius and $P = 0.500656$ days, the mass of the primary must be at least $10^5\ M_\odot$ and the companion must possess an orbital velocity of the order of 10^4 km/sec. Both of these figures are improbably high.

According to de Groot (1969a, 1969b), P Cygni is a very large, high-luminosity star surrounded by three shells formed with material expelled from the star. At present, these concentric shells are stationary with respect to the star and give rise to the three observed absorption components in the spectrum. Other material, moving radially outward from the star passes through these shells with an increasing velocity from 100 up to 200 km/sec. The corresponding mass loss from P Cygni has been variously estimated as between 1.5×10^{-5} and $2.5 \times 10^{-5}\ M_\odot$ per year.

Spectrograms examined by de Groot (1969a) have established that the Balmer series, several lines of He I and the most intense lines of Fe III show two shortward displaced absorption components on the nearly undisplaced emission lines. Indeed, there are often three such components in the case of the Balmer lines of H9 and upward with velocities corresponding to about -95, -125 and -210 km/sec.

In addition, the radial velocity of the component of the Balmer lines most displaced to the violet exhibits variations which appear to have a period of $P = 114$ days. The other lines do not show this periodic variation. The relative intensities of the absorption components also seem to vary but in a quite irregular manner.

A small number of P Cygni variables are known which have somewhat lower intrinsic luminosities than the prototype and here we find that the absorption lines are appreciably broadened, this being indicative of a high rate of axial rotation. From this, we are led to infer that such high axial rotations may make a major contribution to the ejection of matter from the stellar surface in these variables.

Returning now to R Aquarii, the final proof of the complexity of this stellar system comes from long exposure photographs of this strange object. These have shown, quite conclusively, that it consists of a red supergiant star of low surface temperature (varying from ~ 1800 to $2700°K$) which is situated in the midst of a peculiar nebulosity with which it is obviously physically connected. There would also appear to be an O-type companion and an expanding circumstellar shell or shells, the latter giving rise to the P Cygni profiles found in the spectrum. Whether the matter which is being continually ejected into this system originates with the O-type star or not is still uncertain. All in all, we have here one of the most complex stellar systems yet discovered.

References

ALEXANDER, TH. and WALLERSTEIN, G. (1967) *Publ. astr. Soc. Pacif.* **79**, 500.
DE GROOT, M. (1969a) *Non-Periodic Phenomena in Variable Stars*, p. 203, Reidel, Dordrecht.
DE GROTT, M. (1969b) *Bull. astr. Inst. Netherlands* **20**, 225.
MAGALASHVILI, N. L. and KHARADZE, E. K. (1967a) *Inf. Bull. Var. Stars*, No. 210.
MAGALASHVILI, N. L. and KHARADZE, E. K. (1967b) *Observatory* **87**, 295.
MAYALL, MRS. M. (1973) *Q. Bull. A.A.V.S.O.*
NIKONOV, V. B. (1936) *Abastumansk astrofiz. Obs. Gora Kanobili Bull.* **1**, 35.
NIKONOV, V. B. (1937) *Ibid.* **2**, 23.

CHAPTER 22

Herbig–Haro objects

AMONG the heavily obscured regions close to, and within, the Orion Nebula, we find certain tiny emission nebulae which, on photographs taken with large instruments, appear to be semi-stellar in nature. The first of these small, nebulous knots of high temperature gas which may represent an early stage of stellar formation were discovered just over two decades ago by Herbig (1948, 1951) and Haro (1950, 1952) and are universally known as Herbig–Haro objects.

At the present time, some forty or so of these objects are known, most being in the constellations of Orion, Taurus and Perseus and all lying on regions that contain large numbers of T Tauri variables.

Here it is important to distinguish between the true Herbig–Haro objects and other very small, nebulous regions near the Orion Nebula which have a similar optical appearance but are spectroscopically quite different. These latter nebulous spots have been studied by Haro (1953) and Mendez (1967) who have shown that they possess either a conventional emission spectrum or one with a strong continuum in the near infrared.

The fact that there is this strong and close association with T Tauri stars does indeed suggest that there may be some physical connection between the two and this is further strengthened by the observation by Herbig (1968) that two of the T Tauri variables in particular, T and HL Tauri, appear to be immersed within nebulous matter that, both visually and spectroscopically, is very like that found in typical Herbig–Haro objects. At the present time, as shown by Ambartsumian (1954), there is very little evidence against the idea that the T Tauri variables represent a later evolutionary stage of the Herbig–Haro objects.

Certainly it appears quite probable that those Herbig–Haro objects which are not associated with any stars represent regions in which a T Tauri variable will eventually appear. There is little doubt that most of these objects show variations in brightness that suggest that some kind of activity is going on. Magman and Schatzman (1965) have calculated that the degree of ionization which is observed can be produced by a proton flux of the order of 10^5 eV.

There is, however, a difficulty associated with this idea of stellar formation within these nebulous regions. One of these objects, Harbig–Haro Object No. 2, has been assiduously followed by Herbig 9(169) using both the Crossley and the 120-in. reflector of the Lick Observatory since 1954.

This object consists of at least ten nuclei and several of these are variable in brightness. One of the most luminous of them has shown an appreciable fading following its maximum of $\sim 16^m.4$ in 1954–55 and obviously, the fact that a region within these objects can fade as well as brighten, indicates that any increase in luminosity it may produce is not necessarily

The Nebular Variables

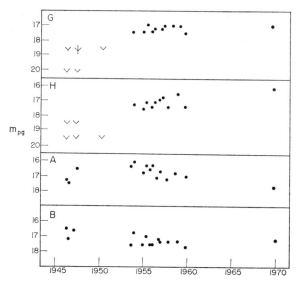

FIG. 54. Variation in photographic magnitude of four of the nuclei in Herbig–Haro Object No. 2.
(After Herbig.)

a permanent feature as it would be in the event of star formation. The light curves of four
of the nuclei in Herbig–Haro Object No. 2, from 1945 to 1968, are illustrated in Fig. 54.

Emission Spectra of the Herbig–Haro Objects

One of the characteristic features of these objects is their emission spectrum which is
unlike that of most of the other nebulous objects that have been discovered. Indeed, this
has enabled astronomers to readily differentiate between the Herbig–Haro objects and the
nebulous knots of gas near the Orion Nebula discovered by Haro and Mendez.

Three of these objects (Nos. 1, 2 and 3) are located in a peculiar region close to the small
nebula HGC 1999, a region which itself is abnormal both visibly and spectroscopically.

The spectral features which are indicative of a Herbig–Haro object have been defined by
Herbig (1969) and shown to be as follows: Strong emission lines of H I and [N II], and
intense lines of [O I] and [S II]. In those instances where the objects are not heavily reddened,
we also find emission lines of [O II].

HERBIG–HARO OBJECT NO. 1

This is the brightest of these objects and consequently it is the one whose spectrum has
been studied in most detail since it is possible to examine several of the less intense emission
lines. Herbig (1969) has shown that the spectrum contains, in addition to the lines given
above, the line of Mg I at 4571 Å, the H and K lines of Ca II and the infrared lines of
[Ca II], [Fe II] and [Fe III]. These lines, in particular, serve to distinguish such objects from
the ordinary gaseous nebula in which they are either absent altogether or exceptionally weak.

More recently, Bohm et al. (1972, 1973) have carried out a spectrophotometric study of
this object using spectra with a dispersion of 51 Å/mm in the blue and 134 Å/mm in the
red, obtained with the Cassegrain image-tube spectrograph of the 84-in. reflector at Kitt
Peak Observatory. New spectral line identifications include additional lines of [Fe II] and

148

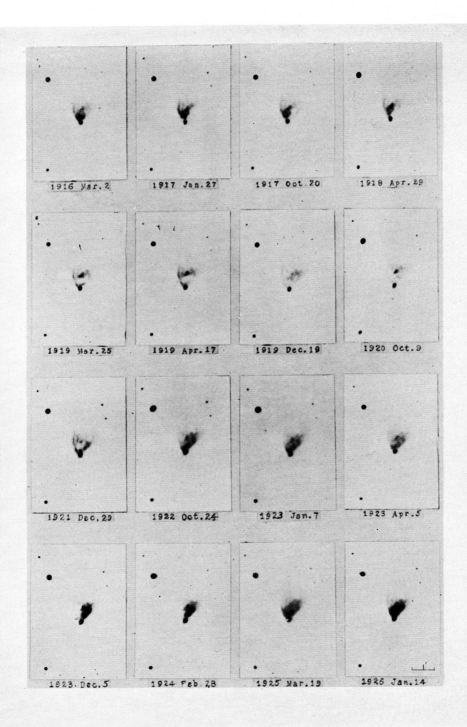

PLATE VI. Object Herbig No. 2 photographed with the 120-inch reflector in red light on December 6, 1959 showing the individual nuclei. (Lick Observatory Photograph.)

[N II] as well as the Balmer lines of H10, H11 and H12. The radial velocity of the object has been determined from these spectra -3.2 ± 6.7 km/sec.

Infrared spectra out to 3μ show no detectable emission at this wavelength. In this respect these objects differ from the T Tauri stars which show a large infrared excess.

HERBIG–HARO OBJECT NO. 2

Being appreciably fainter than the preceding object, the weaker emission lines have not been identified with the same degree of certainty. Herbig (1969) has made infrared observations of this object out to 0.8μ, but these show only the same structure as in the emission lines in the red and blue spectral regions.

One important fact has been pointed out by Herbig (1969), namely that the T Tauri variables themselves show such nebular lines in integrated light. In addition, the low temperature continuum is strongly reminiscent of the T Tauri stars. Here again, we have this close correlation between these variables and the Herbig–Haro Objects.

The general properties of the intense emission lines in the spectra of these objects have been examined by Bohm (1956) and also by Osterbrock (1958) who have suggested that they, and the fact that their blue colour is due to continuous emission and not high temperature, can best be explained on the assumption that both hydrogen and oxygen are not completely ionized under the conditions prevailing in these nebulous regions.

Since these elements are only partially ionized throughout the entire volume of the nebula, it appears that we can exclude the possibility of a single internal source producing radiative ionization. A more probable explanation is that of ionization by high energy particles. The proton flux required to give the observed degree of ionization in Herbig–Haro Object No. 1, for example, is ~ 100 KeV as found by Magman and Schatzman (1965), assuming that $T_e \sim 7500°$K and $n_e \sim 3 \times 10^3$ cm^{-3}.

Structure of the Herbig–Haro Objects

From the photographs which are available it is quite clear that, in the majority of cases, the structure of these objects is comparatively simple. High resolution plates taken with large instruments show that they are not stars and several instances are known where the central condensation is intimately associated with a wisp of nebulous material rather like the tail of a comet. It is perhaps significant that such nebulous appendages have been found around several T Tauri and related variables.

A small number, however, are found to possess a much more complex structure which consists of several individual nuclei. We have already mentioned Herbig–Haro Object No. 2 which contains ten such nuclei. Some of these have remained essentially constant in brightness, while others have, over the years, shown appreciable variation.

The earliest photographic evidence we have of this particular object is on wide-angle photographs taken by Wolf (1927) in 1901 and, although it is far from easy to make reliable estimates of its brightness then, it appears to have been of approximately the same total magnitude as it is at present. This, however, is not particularly surprising since recent photographic evidence shows that the individual nuclei both brighten and fade and such random fluctuations in brightness will tend to nullify each other.

The Nature of the Herbig–Haro Objects

Apart from the generally accepted fact that the Herbig–Haro objects are young formations

The Nebular Variables

it is not easy to put forward a theory that will account for all of the observations, particularly of such a complex object as Herbig–Haro Object No. 2. While several hypotheses have been advanced there is still no straightforward interpretation of the peculiar light variations which have been observed, nor can we be certain whether variability is the exception or the rule as far as these objects are concerned.

(a) One possibility is that here we have a dense cloud of neutral material which would normally be extremely faint, or even invisible, on photographs. If, now, there happens to be a source of ionizing radiation concealed within the cloud, a fairly abrupt stream of such radiation could ionize certain parts of the cloud with the formation of visible, semi-stellar nuclei such as are found in the Herbig–Haro objects. This would particularly be the case for Herbig–Haro Object No. 2 where several such nuclei have been observed to form and brighten over a relatively short period.

On this basis, the decrease in brightness of certain nuclei can be explained by the recombination of the ionized atoms with free electrons subsequent upon the removal of the ionizing source.

Now from the estimated temperature of the nebula which is about 7500°K, and its calculated density, Aller (1956) has shown that the time scale necessary to decrease the surface luminosity, as measured by the hydrogen emission lines, by $0^m.5$ or so in accordance with the observed decrease in brightness, is of the order of 5 years.

A point in favour of this hypothesis is that for the brightest of the nuclei in Herbig–Haro Object No. 2, this is a measure of the observed time scale. However, there still remains one objection to the idea of the ionized hydrogen recombining on such a short time scale as 5 years. Such an ionizing source would have to be concealed within all of the Herbig–Haro objects. In addition it would have to continually irradiate the surrounding nebulous region otherwise they, too, would fade considerably in brightness over a very short period and such fadings have not been observed.

A further piece of evidence against this theory is that we would expect any new nuclei which are formed to possess a different degree of ionization to that of those which have been present for a relatively long period. Again, this does not agree with observation since it appears that all nuclei have the same degree of ionization.

Furthermore, as we saw earlier, the emission lines in the spectra of these objects can be explained on the assumption that the hydrogen and oxygen are only partially ionized throughout the whole volume of the nebulous region. It is very difficult, if not impossible, to explain this on the basis of a single ionizing source situated within the object.

(b) A second alternative, that ionization brought about by particles having a high energy, of the order of 100 KeV as calculated by Magman and Schatzman (1965), could produce this overall partial ionization throughout the entire nebula is possible but again, it would mean that every Herbig–Haro object is irradiated by particles having just this particular energy. At the moment it is difficult to see why this should be so.

The appearance of nuclei G and H in 1954 very close to the previously known nuclei A and B had led to a further suggestion which we must now consider.

(c) The fact that the two new nuclei lie very close to others which have been known for some time implies that, perhaps, orbital motion about a common centre of gravity in this object can explain the observed variations in brightness. Prior to 1954, nuclei G and H may have been in eclipse behind A and B. The gradual brightening would then be due to their emergence from behind these two nuclei. The same mechanism will explain the observed fadings of other nuclei.

Herbig (1969) has shown this theory to be untenable for the following reasons.

(1) If there is any revolution around a common centre of gravity, then the positions of nuclei A and B, taken with respect to a reference frame provided by nearby stars, should have changed over the period when the two new nuclei appeared. Very careful measurements of the available plates have shown that, within the limits of errors of measurement, nuclei A and B occupy the same positions now as they did prior to 1954.

(2) If orbital motion is a contributory factor, there would be a period, prior to 1954, when nuclei G and H were present but unresolved due to their proximity to A and B. Nuclei A and B, therefore, taken effectively as (A+H) and (B+G) would then have been approximately $0^m.75$ brighter than they are now. Such a brightening was not, however, observed.

(3) Both nuclei A and H have subsequently been found to vary in brightness indicating, quite clearly, that individual nuclei are intrinsically variable.

On these grounds there is no evidence at all of orbital motions within this object.

(d) We must now examine the idea that there might be a variable star, possibly of the T Tauri type, embedded within each nucleus, this giving rise to the observed variations in brightness. There are, of course, precedents for this, in that both T and HL Tauri are found in small nebulae of the Herbig–Haro type. It is also true that these variables appear to have, at times, a profound influence upon the luminosity of their parent nebulae. An example of this behaviour is RY Tauri (Herbig, 1961).

The main objection to this hypothesis comes from an examination of the spectra of the Herbig–Haro objects. In the red region especially, the greater portion of the energy is present, not in the continuum as it would be if a variable star were the source, but in the emission lines. We also have the evidence of high resolution photographs and spectrograms taken between 5200 and 5800 Å and in the region from 6800 to 8800 Å which clearly show the nuclei to be non-stellar in these regions of the spectrum.

(e) So far, we have proceeded on the assumption that the variations in brightness of the Herbig–Haro objects are due to some intrinsic property of the nebulous nuclei themselves. It is, however, possible that their light is essentially constant and the changes which have been observed are merely due to periodic veiling by some obscuring dust and gas cloud. Certainly, these objects are all found in regions that are heavily obscured and this argument is worthy of serious attention.

In order to determine its plausibility, we must examine the velocity required of such clouds if their motion is to explain the rather rapid brightening of one or two nuclei as found in Herbig–Haro Object No. 2. These individual nuclei rose from $\sim 20^m.0$ to $17^m.0$ in a period of not more than 3 years.

From evidence we have of other dust regions near the Orion Nebula it seems that, in general, these have motions consistent with a velocity of ~ 1–2 km/sec. Now what would be the angular motion of such a cloud at the distance of this particular object, namely ~ 500 pc?

A simple calculation shows this to be about $0''.5$ in a millenium. The nuclei in this Herbig–Haro object have diameters of the order of $1''.0$ and clearly, a dust cloud moving with the velocity just mentioned would take some 2000 years to cover this angular distance. The velocities necessary to explain the rapid increases in brightness that have been observed are, therefore, impossibly high.

(f) One further possibility has been put forward by Herbig (1969); this being that the complex Herbig–Haro objects which consist of several variable nuclei are, in reality, dense

The Nebular Variables

clouds of dark material which are gravitationally coherent. The observed light-variable nuclei are then, not permanent features, but rather transient phenomena occurring on the surface of such a cloud in a manner somewhat analogous to sunspots. It is even possible that the changes which have been observed in the brightness of the nuclei over the past few years may be due to some cyclical phenomenon.

Whether or not the cloud is gravitationally stable will depend upon the mean density. Herbig has demonstrated that if the whole region containing the bright structures within Herbig–Haro Object No. 2 has a mean hydrogen density of $N \approx 10^4$ cm^{-3}, the total mass would be only about 0.05 $M_\odot$ with a surface escape velocity of only ~ 0.1 km/sec. Under these conditions, the structure would not be gravitationally bound. It appears more probable, however, that the mean hydrogen density is appreciably higher than this with $N \approx 10^6$ cm^{-3} and in this case, the cloud could remain essentially intact.

Whatever the real nature of these objects might be, it seems clear at the moment that a great deal of observational work with large instruments is still necessary before we arrive at the correct picture. It is especially desirable that some indication should be gained concerning any variations which take place in the fine structure of the nuclei whenever a brightening or fading occurs. One further point which still requires clarification is whether a particular nucleus can fade or brighten several times or only once.

References

ALLER, L. H. (1956) *Gaseous Nebulae*, p. 66, Wiley, New York.
AMBARTSUMIAN, V. A. (1954) *Mem. Soc. Roy. Sci. Liége*, **14**, 293.
BOHM, K. H. (1956) *Astrophys. J.* **123**, 379.
BOHM, K. H., PERRY, J. and SCHWARTZ, R. (1972) *Bull. Am. astr. Soc.* **4**, 243.
BOHM, K. H., PERRY, J. and SCHWARTZ, R. (1973) *Astrophys. J.* **179**, 149.
HARO, G. (1950) *Astron. J.* **55**, 72.
HARO, G. (1952) *Astrophys. J.* **115**, 572.
HARO, G. (1953) *Ibid.* **117**, 73.
HERBIG, G. H. (1948) *Thesis*, Univ. of California, California.
HERBIG, G. H. (1951) *Astrophys. J.* **113**, 697.
HERBIG, G. H. (1961) *Ibid.* **133**, 337.
HERBIG, G. H. (1968) *Liége Astrophys. Symposium.*
HERBIG, G. H. (1969) *Non-Periodic Phenomena in Variable Stars*, 75, Reidel, Dordrecht.
MAGMAN, C. and SCHATZMAN, E. (1965) *Seanc. Acad. Paris* **260**, 6289.
MENDEZ, M. E. (1967) *Bol. Obs. Tonantzintla Tacubaya* **4**, 104.
OSTERBROCK, D. E. (1958) *Publ. astr. Soc. Pacif.* **70**, 399.
ROBERTS, I. (1927) *Isaac Roberts' Atlas of 52 Regions*, Chart 22.

CHAPTER 23

Symbiotic variables

AMONG the irregular variables we find a small number which, in addition to their erratic and unpredictable variations in brightness, also possess very curious spectra that are somewhat similar to that which was discussed earlier for R Aquarii. The term "symbiotic variables" was first given to these stars by Merrill (1950) and is now widely used for a variety of objects which have composite spectra consisting of absorption features such as are produced by a low-temperature giant star, together with emission lines of high excitation. The latter indicate the presence of a high-temperature component. In addition, nebulous material is also present in these complex systems which contributes, in part at least, to the observed variations in brightness.

In a general survey, Bidelman (1954) published a list of 23 stars having such composite spectra while Payne-Gaposchkin (1957) put forward evidence for the inclusion of 32 stars in this class. As might be expected, most of the stars given are common to both lists.

As more data accumulated, it became clear that the presence of a composite spectrum is not, in itself, a sufficiently precise criterion for defining a star as a genuine member of the symbiotic class. Some of the long period variables, notably R Aquarii, also satisfy such a criterion. This also applies to certain of the recurrent novae such as T Coronae Borealis and RS Ophiuchi. Indeed, according to this definition, the dwarf novae could be classed as symbiotic variables.

From detailed investigations of several typical symbiotic stars, Boyarchuk (1969) has proposed that such variables must satisfy the following criteria.

(a) The spectrum must show the absorption lines and bands of a late spectral type; for example, bands due to TiO or ZrO and lines of Ca I and Ca II.

(b) There must be emission lines of ionized atoms present such as H I, He I, [O III], [Ne III], [S II] and [A III].

(c) The width of these emission lines must be small, not exceeding a value of ~ 100 km/sec.

(d) The brightness of the object must be variable. Here we find that the star usually varies with an amplitude of anything up to 3^m in a period of several years. In addition, the light curve shows the variations to be irregular with the ascending branch steeper than the decline. Moreover, during the decline in brightness, both permitted and forbidden lines of progressively higher excitation and ionization develop.

Table XVIII contains the available data on the symbiotic variables known at the present time, while Table XIX lists those variables for which complete data are not available but which are probable members of this class.

As will be seen, the majority of these stars are quite bright, particularly around maximum light, and are readily visible in small instruments during this phase of their light variation.

The Nebular Variables

Light curves for several of them have been produced by the American Association of Variable Star Observers (1963).

TABLE XVIII. SYMBIOTIC VARIABLES

Star	Magnitude Maximum	Magnitude Minimum	Spectrum	Major emission feature	Reference
Z And	8.0	12.4	M2 III	[Fe VII]	Boyarchuk (1967a)
CM Aql	13.2	16.5	M4 III	He II	Herbig (1960)
BF Cyg	9.3	13.5	M5 III	[O III]	Boyarchuk (1968a)
CH Cyg	6.6	7.8	M6 III	[Fe II]	
CI Cyg	9.4	13.7	M5 III	[Fe VII]	Boyarchuk (1968a)
V407 Cyg	13.3	16.5	Mep	—	Merrill, Burwell (1950)
V1016 Cyg	10.0	15.5	M3 III	[O III]	Boyarchuk (1968b)
AG Dra	9.1	11.2	K3 III	He II	Boyarchuk (1966)
YY Her	11.7	13.2	M2 III	[O III]	Herbig (1950)
V443 Her	12.4	12.9	M3 III	[O III]	Tift, Greenstein (1958)
RW Hya	10.0	11.2	M2 III	[O III]	Merrill (1950)
SY Mus	11.3	12.3	M3 III	[O III]	Heinze (1952)
AR Pav	10.2	12.7	M	[O III]	Sahade (1949)
AG Peg	6.8	8.2	M3 III	[O III]	Boyarchuk (1967a)
AX Per	9.7	13.4	M5 III	[Fe VII]	Boyarchuk (1968a)
RX Pup	11.1	14.1	M5 III	[Fe VII]	Swings, Struve (1941b)
FN Sgr	9.0	13.9	Pec	[O III]	Herbig (1950)
FR Sct	11.7	12.5	M2 III	[O III]	Bidelman, Stephenson (1956)

TABLE XIX. PROBABLE SYMBIOTIC VARIABLES

Star	Magnitude Maximum	Magnitude Minimum	Spectrum	Major emission feature	Reference
DV Aur	8.2	10.0	C5	[O III]	Sanford (1944)
Y CrA	12.0	12.9	Pec	—	Bidelman (1954)
HK Sco	13.1	15.8	Pec	He II	Elvey (1941)
KW Sco	11.0	13.2	Mp	—	Swope (1940)
V455 Sco	12.8	16.5	Composite	H	Merrill, Burwell (1950)
HD 4174	7.5	—	M2 III	[O III]	Wilson (1950)
HD 330 036	11.7	—	—	[O III]	Webster (1966)
Hz 134	—	15.0	—	[O III]	Webster (1966)
Hz 172	—	12.9	—	[O III]	Webster (1966)
F 6–7	11.0	—	M4	H	Merrill, Burwell (1950)
MHα79–52	12.0	—	—	[Fe VII]	Merrill, Burwell (1950)
MHα80–5	11	—	Composite	H	Merrill, Burwell (1950)
MHα276–12	—	17	—	[Fe VII]	Merrill, Burwell (1950)
MHα276–52	11.5	14.5	—	[Fe VII]	Merrill, Burwell (1950)
MHα305–6	11.5	—	—	[Fe VII]	Merrill, Burwell (1950)
MHα359–110	11	—	—	[Fe VII]	Merrill, Burwell (1950)

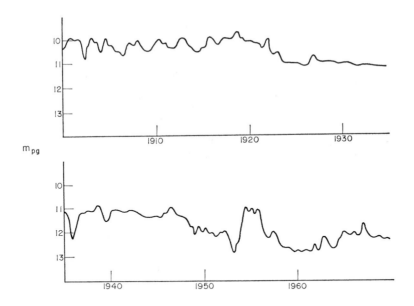

Fig. 55. Photographic light curve of BF Cygni.

Light Variations of the Symbiotic Variables

All of the symbiotic variables, with the exception of those stars such as R Aquarii which are also long period variables, exhibit totally irregular fluctuations in brightness. Very often there is the impression of flare-like activity, while at other times the brightness remains essentially constant for quite long periods. This behaviour is readily seen from the light curve of BF Cygni (Fig. 55).

Although the majority of the symbiotic variables show similar light variations, it is often necessary to treat each star individually since not only do different stars possess somewhat different light curves, but even in the case of the same star we find periods when the light variations change in character, sometimes rather abruptly.

Broadly speaking, however, most of these variables have light curves which closely resemble that of BF Cygni, characterized by four types of variation.

(a) There are periods when the star changes little in brightness, such periods often lasting for several years.

(b) At times, there are semi-periodic fluctuations in brightness which have amplitudes of up to 1^m.

(c) At intervals, the brightness varies in a completely irregular manner, often quite rapidly, suggestive of minor flares.

(d) Occasionally, there are major flares such as those shown by Z Andromedae in 1914 and 1939 with amplitudes which can be as much as 4^m (Fig. 56).

A few of these stars exhibit light variations that are completely unlike those of the majority of the variables in this class. AG Pegasi, for example, shows a very peculiar variation which, in some respects, resembles that of the slow nova η Carinae. Visual observations of this star first began about 1825 but unfortunately these were extremely sporadic and only serve to delineate the general trend rather than define any short-term changes. These early

155

observations have been summarized by several authors including Lundmark (1921), Himpel (1942), Sandig (1950) and Payne-Gaposchkin (1950).

It is only comparatively recently that more frequent and detailed observations of AG Pegasi have been made, such as those published by Mayall (1964) and Belyakina (1965). From these, the latter has demonstrated that there is a pseudo-periodic variation in brightness which is sinusoidal in character with an amplitude of $0^m.4$ and with a mean cycle length of ~ 800 days.

So far, the fluctuations of these stars which we have been discussing have been concerned with the long-term changes. It must not, however, be thought that there are no shorter-term variations superimposed upon these. In particular, Belyakina (1965, 1967, 1970a, 1970b) has carried out accurate photoelectric measurements on certain of the symbiotic

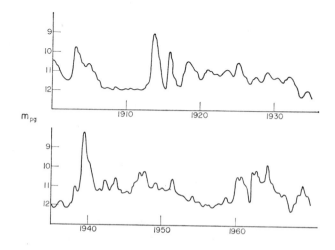

Fig. 56. Photographic light curve of Z Andromedae.

variables with some very interesting results. As we might expect, the rapid changes in brightness revealed by the photoelectric cell have only small amplitudes and exhibit no periodicity at all.

Three colour photometry was used in this work and we are therefore able to reach some important conclusions regarding the cause of these rapid fluctuations. AG Pegasi is one of the stars examined and observations made from 1962 to 1967 have shown that the light changes are synchronous in all regions of the visible spectrum and also in the ultraviolet. In the yellow and blue regions the amplitude is $0^m.3$, whereas in the ultraviolet it is $0^m.5$.

These photoelectric measurements also show that as these stars decrease in overall brightness, the ultraviolet excess increases. This provides confirmatory evidence for the earlier work by Jacchia (1941), Himpel (1940) and Payne-Gaposchkin (1946) demonstrating that the colour of the symbiotic variables changes in step with the visible brightness. As the stars fade, they become redder.

It has been suggested by Belyakina (1970a) that these changes are due to the orbital motion of a cool, giant star, one hemisphere of which faces a hot companion and thereby exhibits a reflection effect. As we shall now see, most of the accompanying spectroscopic changes are in full agreement with such a model.

Spectroscopic Changes in the Symbiotic Variables

As with all variable stars, there are quite considerable variations in the spectra of the symbiotic variables which, broadly speaking, keep in step with their light variations.

The complex features of these spectra have been known for some time and many years ago Hogg (1934) noted that the broad bands of TiO of the late-type absorption spectrum increase in intensity with a decrease in brightness. In addition, there are changes in the emission spectrum corresponding to an increase in the excitation level. Further observations have shown that similar changes in the spectral features in nearly all of the symbiotic variables take place.

Among the emission lines, we often find that the intensities of the He II and [Fe VII] lines decrease appreciably in comparison with those of hydrogen with an increase in stellar brightness. At the same time, the intensity of the TiO bands strongly decreases.

During the period of a major flare, drastic changes occur in the spectra of these variables. Late-type absorption features, together with the highly excited emission lines, disappear altogether. In their place, the absorption lines normally found in stars of spectral types AO to A2 appear while the hydrogen and He I emission lines have absorption components.

The energy distribution in the continuum, as shown by Boyarchuk (1969), tends to become steeper and the Balmer jump increases as the stars fade. The displacement of different absorption lines enables radial velocities to be determined for several of these variables, providing additional proof of orbital motion about a common centre of gravity.

Infrared Excesses of the Symbiotic Variables

The majority of the symbiotic variables have near infrared continua which closely resemble those of late type stars of the Mira class. RX Puppis and V1016 Cygni have been found by Swings and Allen (1972) to possess very prominent infrared excesses which are attributable to radiation from a circumstellar dust cloud. The Bep star MWC 56, whose spectrum shows Fe II and [Fe II] emission lines in the blue, also possesses the absorption lines of a G or K type component similar to the genuine symbiotic variables. Here, however, no substantial infrared excess is present.

Z ANDROMEDAE

As we have already mentioned, Z Andromedae increased in brightness by almost 4^m during the bright, major flares of 1914 and 1939 and several spectrograms of this star were obtained during the period of the latter flare, mainly by Swings and Struve (1941a). These show that many of the spectral features underwent marked changes at this time, all of them being the reverse of those just discussed for the fadings of these stars.

The emission lines due to excited atoms disappeared entirely and the broad absorption bands of TiO were scarcely visible. In place of the latter, several absorption lines were evident similar to those in stars of spectral type A2. One important feature noted at this time was that the emission lines of H I and He I possessed absorption satellites on their violet borders. This is, again, the sort of situation we find in the P Cygni variables and is indicative of an expanding envelope and continuous ejection of material into the surrounding medium.

There is now ample evidence that these changes occur in all of these variables during the time of a major increase in brightness. They were observed during a less spectacular flare of Z Andromedae in 1961 by Bloch (1962) and also in the spectrum of AX Persei during 1955 by Gauzit (1955).

A recent comprehensive study of the spectral variations taking place in Z Andromedae has been made by Bloch *et al.* (1969) covering the period from 1923 to 1968. This has demonstrated that there is no periodicity in either the appearance of lines of different ionization or degrees of excitation.

From a comparison of the Fe II permitted lines at different wavelengths in the spectrum of this variable, Caputo (1971) has derived a colour excess of $E_{B-V} = 0^m.64 \pm 0^m.20$. The curve of growth of the Fe II emission lines indicates that self-absorption is negligible in this system.

CH CYGNI

This star has been classed as a semi-regular variable with a spectral type of M6 by Kukarkin and Parenago (1948) but it appears to be, almost certainly, a member of the symbiotic class, mainly on the basis of its spectrum. Photoelectric observations of the star have been made by Cester (1969) during the outbursts of 1967 and 1968 which reveal marked variability, particularly in the ultraviolet. From the observed colours, it appears that the companion is a hot subdwarf.

Spectrograms taken with a moderate dispersion by Nguyen-n-Doan (1970) following the outburst of 1967, together with ultraviolet spectra, show that the emission lines visible during the outburst had vanished by the following year giving place to an absorption spectrum composed mainly of Ti II lines from the ground state, multiplet 1 to multiplet 8 with other low excitation lines due to Cr II and V II.

At this stage, the Fe II lines were very weak. Late spectra taken by the same observer in 1969 show the presence of a calcium envelope. In the green and red regions, apart from the normal M6 type spectrum, there were also the nebular emission lines of [O I], [S II] and the chromospheric lines of Na I, He I, Fe II and [Fe II]. Quite clearly, following the outburst, the spectrum was a composite one of stellar, chromospheric and nebular components but, in contrast to most symbiotic variables, the excitation potential was found to be rather low.

Faraggiana (1969a) also found strong negative radial velocities of the Ca II chromospheric absorptions and a broadening of the emission lines of H I, He I, Fe II and [Fe II], showing that turbulent motions are present in this envelope. The mass loss from this star has been estimated as $\sim 10^{-8} M_\odot$ per year. Furthermore, Faraggiana and Hack (1969b), from a good quality high dispersion spectrogram (9.7 Å/mm), found a strong similarity with VV Cephei out of eclipse and also some similarity with η Carinae. From these observations there appears little doubt that CH Cygni is a genuine symbiotic variable.

CI CYGNI

The light variations of this variable have been followed by several workers, including Greenstein (1937), Aller (1954) and Hoffleit (1968). In general, the amplitude of this star is quite small ($0^m.5$), but three major flares have been observed in 1937, 1971 (amplitudes $\sim 3^m$) and an earlier one in 1911 (amplitude $\sim 2^m$). Quasi-periodic variations with a cycle length of 815^d have been found by Mrs Hoffleit. From the overall scatter found in the light curve of this variable it is evident that non-periodic variations (probably rapid) are important.

The light curve of the star during the most recent flare of 1971–72 has been published by

Stiemon (1973) who has shown that the activity of the star at this time was comparable with that of the 1911 outburst.

V1016 CYGNI

The peculiar variable MHα 328–116 (V1016 Cygni) was recognized as a possible member of the class of symbiotic variables when Merrill and Burwell (1950) discovered that although the spectrum is predominantly of type M, the Balmer line of Hα appears bright in emission. However, it was only in 1965 that its peculiar nature was fully recognized when McCuskey (1965) showed that it had brightened by more than 3^m while at the same time its spectrum contained a large number of emission lines. Over the past few years this star has continued to brighten slowly until it is now $\sim 10^m.0$.

Since 1966, the spectrum of this object has changed remarkably and a complete examination of these variations has been carried out by Mammano and Rosino (1966, 1968, 1969) using the 56-in. telescope of Asiago Observatory. In general, the degree of excitation of the emission lines increased dramatically, especially those of He II (4686 Å). There was also an emergence of forbidden lines of ions with a high ionization potential such as [Ne IV], [Ne V], [A VI], [A V], [Fe V] and of the permitted lines of Fe II.

As Mammano and Rosino have suggested, the most conspicuous changes in the spectrum may be attributed to both stratification of the atoms emitting the lines of different ionization and to variations in the temperature of the hot source in the system associated, too, with a decrease in the density of the surrounding envelope brought about by expansion.

It would seem, therefore, that beginning in 1965, there was a gradual brightening of a high-temperature component in this system, the radiation from which contributed to the ionization of surrounding gases, thereby giving rise to the increasing intensity of the spectral emission lines. At the same time, this brought about an expansion of the envelope with a resulting decrease in its mean density.

A similar conclusion was reached independently by FitzGerald and Houk (1970) from a study of some 130 emission lines and their intensities between 3130 and 5030 Å on spectrograms with dispersions ranging from 12 to 130 Å/mm taken during 1965–68. The line intensities and ratios do not give a completely consistent picture but do indicate the increasing temperature of the emitting regions and the decreasing density. Radial velocities of -60 ± 15 km/sec and line widths of 80–120 km/sec were determined. Again, it may be that this object is a planetary nebula in its early stages of evolution. This hypothesis was also put forward by Mammano and Rosino (1968).

Further observations by Philip (1969), Bloch (1969) and Boyarchuk (1968b) have essentially confirmed these observations. The latter has shown that the relative intensities of the emission lines and the energy distribution in the continuum may be explained on the basis of a binary system embedded within nebulosity. One component is an M type giant and the other a hot star with a surface temperature of $\sim 10^5$ °K. The surrounding nebula has an electron density of $n_e = 2.5 \times 10^6$ cm^{-3}, an effective temperature $T_e = 18,000$°K and a mass of 1.55 M.

AG DRACONIS

This symbiotic variable has been comprehensively observed by Belyakina (1969) using three colour photometry similar to the UBV system. The combined photoelectric light curve covering the period from 1062 to 1067 shows that the light fluctuations are irregular. They are, however, synchronous in all regions of the spectrum (like those found for AG

159

The Nebular Variables

Pegasi mentioned earlier), with amplitudes of $0^m.15$ in the yellow, $0^m.30$ in the blue and $1^m.0$ in the ultraviolet.

From these results, Belyakina has postulated that the system consists of a cool star (spectral type K5 III), a hot component with a surface temperature of $50,000°K$ and a surrounding nebula with an electron density $n_e = 10^6 \text{ cm}^{-3}$ and $T_e = 17,000°K$. It would thus appear that the light variations in all regions of the spectrum are due to the variability of the hot companion. The amplitude of the light variations of the hot component in yellow light exceeds $2^m.5$.

AG Pegasi

The light variations of this variable are of a different character altogether to those of other symbiotic stars. Prior to 1842, the star had been essentially constant at $9^m.1$ and then commenced a slow brightening, attaining $6^m.2$ about 1870. Almost at once, however, it began a more gradual decline to its present brightness of $\sim 8^m$. Recent photoelectric and visual observations show the semi-periodic variation between $8^m.0$ and $8^m.3$ with a cycle time of $\sim 800^d$.

Hutchings and Redman (1972) have reported results of velocity measurements and spectrophotometry of this star during 1970–71. Various phases in the periodic variation of the spectrum have been noted. A variable Balmer emission line progression has also been discovered in all spectra. About 70 absorption lines due to the M type component have been identified.

AX Persei

The light variations of this star are highly irregular and have been discussed by Lindsay (1932) and Payne-Gaposchkin (1946). Others observers are Wenzel (1956), Sieder (1956) and Romano (1960). In general, the light curve is very similar to that of Z Andromedae (Fig. 56).

The changes found in the spectrum of this star by Gauzit (1955) during the major flare of that year are also almost identical with those of Z Andromedae in flare.

RX Puppis

This southern symbiotic variable has a light curve somewhat like that of CI Cygni. Swings and Struve (1941b) have shown that this similarity also extends to their spectra. That of RX Puppis shows, besides strong emission lines of H I, He I, He II, strong lines of [Fe VII] and other, less intense lines of [Ne V], [Fe Vi] and [Ca VII]. There is, however, less evidence for a late type component in this system.

HBV-475

The peculiar emission variable star Hamburg–Bergedorf Variable 475 (HBV-475) has several photometric and spectral characteristics similar to those of V1016 Cygni. The photographic light curve from 1891 to 1970 has been studied by Shao (1971). This reveals the following phenomena:

(a) Between 1891 and 1965, the mean brightness of this object was $m_{pg} \sim 15^m$ with fluctuations of the order of $\pm 1^m.0$.

(b) During this interval, several deep minima to as faint as 18^m were observed with a suggested period of 2500^d. Unfortunately, a lack of sufficient data makes it difficult to determine a precise period.

(c) The abrupt nova-like outburst during the winter of 1965–66 was very unusual, although somewhat reminiscent of V1016 Cygni. The star has remained near maximum brightness ever since.

(d) While the light curve is not unlike those of typical symbiotic variables, the amplitude of more than $6^m.5$ is the largest of all these objects.

HV 13055

This interesting variable is a member of the Lesser Magellanic Cloud and, according to Hodge and Wright (1970), has a light curve, amplitude, cycle length and colour index very similar to those of the symbiotic variables. The absolute bolometric magnitude of $M_b = -3^m.2$ at maximum and $M_b = -0^m.4$ at minimum agrees well with the available spectroscopic data for galactic symbiotic variables and indicates that the cool component of this particular system is a red giant.

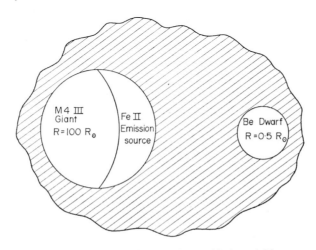

Fig. 57. Schematic diagram of a symbiotic variable.

At present, this is the only probable symbiotic variable to have been positively identified in an extragalactic system and clearly, further spectrophotometric observations of this object are highly desirable.

The Physical Nature of the Symbiotic Variables

From the foregoing discussion of representative members of this class it will have become apparent that these systems are both peculiar and complex in their structure.

From their spectra alone, it would appear that there are three sources of radiation present; namely a low-temperature star that provides the main absorption features similar to those found in the long period variable and semi-regular stars, a high-temperature source which produces the varying degree of exciting radiation found in the emission lines and a surrounding nebula.

This picture is shown schematically in Fig. 57.

One method of determining the presence of a binary system is to study the radial velocity curves for the various spectral lines since, if these exhibit periodicity, we have quite good evidence for duplicity. In the case of R Aquarii, the radial velocity curve has a periodic

The Nebular Variables

character although there is a certain degree of skewness indicating either an elliptical orbit or interference from the surrounding nebulosity.

Z Andromedae, BF Cygni and AG Pegasi have been shown by Swings and Struve (1943) to possess periodic variations in their radial velocity curves (Fig. 58) which, as can be seen, are symmetrical.

From the calculations made by Boyarchuk (1969) on the variations in brightness of the three probable components of Z Andromedae during the period from 1960 to 1965, it would seem that whereas the magnitude of the surrounding nebula and the hot component changes appreciably, that of the cool star varies by only a small amount. A good deal of data have now accumulated regarding the colours and spectral characteristics of the symbiotic variables. Investigations of the energy distribution in the underlying continuum have been carried out by several workers in this field including Tcheng Mao Lin and Bloch

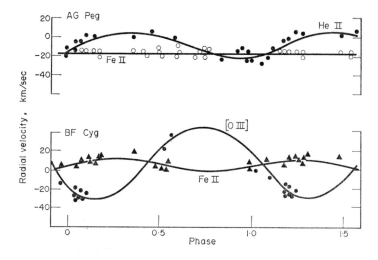

Fig. 58. Radial velocity curves for AG Pegasi and BF Cygni. (After Boyarchuk.)

(1952, 1954), Boyarchuk (1967a) and Ivanova (1960), all of which show that as the star brightens, the maximum intensity in the continuum moves towards the violet, the reverse occurring as the brightness declines.

The physical nature of the variations in the hot component to produce these changes is still not fully resolved although two possibilities suggest themselves. The formation of an optically thick atmosphere around this object could result in the observed light fluctuations, as could some form of pulsation.

Now the cool components of these variables are normal giants and clearly they will have evolved to the right of the main sequence. The hot members, on the other hand, are found to be situated below the main sequence in approximately the same region as the white dwarf companions of the dwarf novae and also the central stars of the planetary nebulae (Fig. 59).

All of these objects are known to possess some form of instability and it is quite conceivable that the same kind of process may be present in the hot stars of the symbiotic variables.

Alternative Theories of the Symbiotic Variables

As we have just seen, there is some spectroscopic evidence that some, if not all, of the symbiotic variables are binary systems. However, certain other possibilities have been suggested in recent years and here we may review some of these in detail.

The idea that these objects may consist of a single star was first put forward by Sobolev (1945) and enlarged upon by Mendel (1946) and Aller (1954). The picture we have according to this theory is that of a high-temperature star which is surrounded by an optically thick envelope. We may thus regard the central star as a curious intermediate stage between a red giant and a blue degenerate phase.

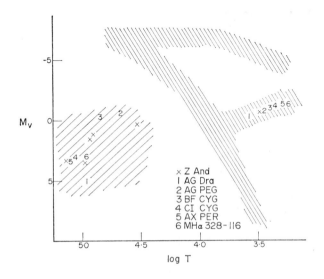

FIG. 59. Positions of the primary and secondary components of typical symbiotic variables on the Hertzsprung–Russell diagram. (After Boyarchuk.)

The absorption features in the spectra of these stars are considered to arise in the outermost regions of this extensive circumstellar cloud which will be at a much lower temperature than that of the central star. It is the latter, and also the inner regions of the envelope, which will be affected by the radiation from the star itself, that are the sources of the emission spectrum. As may be imagined, much of the radiation from such a hot star will be in the ultraviolet and this will have the effect of ionizing the gases nearest the star.

When we come to examine the spectra in detail, however, we find that there are some grave objections to this hypothesis. For example, from theoretical considerations, we would expect the emission lines to vary widely in intensity in a manner analogous to those found in the spectra of the long period variables. Here, the emission lines are at their most intense shortly after maximum light and fade appreciably, and in a regular manner, towards minimum brightness. Other minor fluctuations in the emission lines also occur and these have not been observed in the spectra of the symbiotic variables.

An alternative theory suggested by Aller (1954) and Gauzit (1955) is that of a low-temperature star which is surrounded by a very extensive corona of relatively high density.

163

The Nebular Variables

Now such stars are indeed known in quite large numbers, being found mainly among the semi-regular variables. The third magnitude star α Herculis is believed to be surrounded by a vast circumstellar envelope which is being continually replenished by mass ejected from the star. Fortunately this star is a member of a close binary system, the companion being of spectral type G. Examination of the spectrum of this companion shows that the circumstellar lines are almost as intense as in the M-type component, indicating that the huge gaseous envelope extends at least as far as this component.

Once again, however, we run into serious difficulties in applying this idea of such a system to the symbiotic variables. With such a cool star it is impossible to explain the high excitation of the emission lines since we can find no method of heating the outer photosphere to the point where these lines are produced in the spectrum without raising the temperature of the entire photosphere to this temperature. In other words, there is no known manner by which such a low-temperature star, with a surface temperature of between 1700 and 3000°K, can heat a small region of the photosphere to the required temperature.

The idea of a three-component system is, therefore, the one which is accepted by most astronomers at the present time. The parameters of these components are typically as follows:

(1) The cool component: M or K type giant.
Radius: $\sim 100\ R_\odot$.
Surface temperature: 3000–4000°K.
Absolute visual magnitude: $-0^m.5$.

(2) The hot component: Be type dwarf or subdwarf.
Radius: $\sim 0.5\ R_\odot$.
Surface temperature: 100,000°K.
Absolute visual magnitude: $+0^m.5$.

(3) The surrounding nebula:
Radius: $\sim 50,000\ R_\odot$.
Effective temperature: $\sim 17,000°K$.
Electron density: $5 \times 10^6\ cm^{-3}$.
Mass: $0.0001\ M_\odot$.

While the latter hypothesis explains many of the features of the light curves and spectral variations of the majority of the symbiotic variables there are, as we have seen, a small number of objects, typified by V1016 Cygni, for which an alternative theory seems a definite possibility.

The theory put forward by Aller (1954) and Gauzit (1955) may apply in this case. Quite clearly, a red giant star which is evolving into a blue degenerate phase and accompanied by a high density corona could evolve further into a planetary nebula. More observational data of both light and spectral variations will, however, be required over the next few years before we are able to make a choice among the various hypotheses put forward to explain these peculiar objects.

One possible piece of evidence we have at the moment comes from the work of Boyarchuk (1970) who has shown that, from the relative intensities of the emission lines in the spectra of several symbiotic variables, the chemical composition of the associated nebulae does not differ from that of the planetary nebula NGC 7027.

164

Population II Symbiotic Variables

The red giant components of all the symbiotic variables we have just been discussing are conventional M type stars of Population I. Recently, Herbig (1969) has examined low dispersion slit spectrograms of faint emission objects lying in the direction of the galactic bulge, all of these showing Hα emission upon a detectable continuum.

Eleven of these stars have been identified as certain, or probable, symbiotic variables and comparison with R Coronae Borealis stars in the same region suggests that the brightest symbiotic variables in the galactic bulge have mean absolute visual magnitudes of $M_v \sim$ -3^m to 4^m. This clearly suggests that the cool components of these particular systems are red giants of Population II.

Among the reviews of symbiotic variables which have been published during the last 15 years, we may mention those by Merrill (1968), Payne-Gaposchkin (1957) and Sahade (1960, 1965).

Radio Emission from Symbiotic Variables

Although much of the infrared excesses found in the symbiotic variables (and other early type emission line stars) can be attributed to thermal re-radiation from the circumstellar dust, there is a contribution in some cases from free-free emission by the circumstellar gas as shown by Woolf, Stein and Strittmatter (1970). If this gas happens to be optically thin at centimetre wavelengths, it is to be expected that some of these objects will be sources of radio emission.

A search for radio emission at 10.63 GHz from certain symbiotic variables has been made by Purton, Feldman and Marsh (1973) using the 46-m telescope of the Algonquin Radio Observatory. A positive radio emission was detected from V1016 Cygni of 85 ± 10 m.f.u. Although the random noise level was quite appreciable, the agreement in position with V1016 Cygni made it virtually certain that this emission at 10.63 GHz originates from the variable. Previous attempts to detect radio emission from 1016 Cygni have been quoted by FitzGerald and Houk (1970).

While the near infrared spectrum of this variable can be attributed to thermal radiation from a shell of circumstellar dust with an effective temperature of $T_e \sim 1000°K$, the far infrared spectrum shows the presence of another dust component with a much lower effective temperature of $T_e \sim 300°K$ as suggested by Knacke (1962).

In addition to extending the known spectrum of this object, the radio emission at 10.63 GHz also reveals another property of the variable. The predicted radio emission expected from the black body spectrum of the cooler dust component whose presence is indicated by the far infrared spectrum is considerably lower than that which has been measured. As Purton *et al.* (1973) have pointed out, it seems reasonable to assume that the radio emission comes from free-free processes in the circumstellar gas. In addition, since the absolute level of the radio emission is some two orders of magnitude less than that found in the near infrared, it may be assumed that the Bremsstrahlung radiation from this gas contributes very little to the observed infrared excess.

HBV-475 has also been found to possess a detectable radio emission by Altenhoff and Wendker (1973). In this variable, too, the relative intensities of the radio emission and the infrared component are very similar to those in V1016 Cygni. It seems logical, therefore, that the role of free-free processes in producing the infrared excess in this object is similar in both variables.

The Nebular Variables

The Recurrent Novae

The intervals between eruptions of the recurrent novae are considerably longer than in the case of the genuine symbiotic variables mentioned above and their amplitudes are correspondingly greater. Their light curves, too, are more regular in the sense that, although minor fluctuations are often observed during periods of minimum brightness, their outbursts are more sharply defined.

Their inclusion here is due more to their physical characteristics than to the form of their light variations. With the exception of T Pyxidis, for which there is no evidence of a Be type companion, all of these variables have been shown to be binary systems.

In addition, we find circumstellar shells of material around these systems, much of this gas having been ejected during the recurrent nova outbursts.

There is also a similarity to the symbiotic variables of the Z Andromedae class in that both permitted and forbidden emission lines are present on the blue continuum during minimum light and these are of the same degree of excitation and ionization as those found in the genuine symbiotic variables.

Here we shall discuss two of the recurrent novae—T Coronae Borealis and RS Ophiuchi—in detail, bearing in mind that all of these stars display much the same kind of photometric and spectroscopic evolution during and after a typical outburst. Details of some of these stars are given in Table XX.

TABLE XX. RECURRENT NOVAE

Star	Outbursts	Absolute magnitude		Remarks
		Maximum	Minimum	
T CrB	1866–1946	−8.4	+0.2	Binary
RS Oph	1898–1933–1958–1967	−8.3	−0.8	Binary
T Pyx	1890–1902–1920–1944–1967	−6.4	+0.8	Single
WZ Sge	1913–1946	−7.3	+1.8	Binary
V1017 Sgr	1901–1919	−6.4	+0.7	Binary

T CORONAE BOREALIS

In May 1866, this star brightened abruptly from $\sim 10^m.0$ to $1^m.8$ in a little over 2 days. The decline set in almost immediately, however, with the star fading to $\sim 8^m$ in less than 20 days. A subsidiary rise was recorded approximately 30 days after the initial outburst, this having an amplitude of $\sim 1^m$.

A similar rise occurred in February 1946 with the star reaching third magnitude, the decline following a similar pattern to that of 1866. The fairly close similarity between the two outbursts may be seen from the light curves given in Fig. 60.

As a measure of the steepness of the decline, we may define the parameters T_2 and T_4 as the time intervals from the maximum brightness during which the light faded by 2 and 4 magnitudes respectively. For the two outbursts, these values are approximately:

$$1866: T_2 = 3^d, \quad T_4 = 6^d \qquad 1946: T_2 = 5^d, \quad T_4 = 8^d.$$

The light curve obtained by Petit *et al.* (1946), based upon 14 determinations with a visual wedge photometer, shows a maximum rate of decline of $0^m.48$ per day which may be compared with Schmidt's estimate of $0^m.63$ per day for the 1866 outburst. The companion to the

166

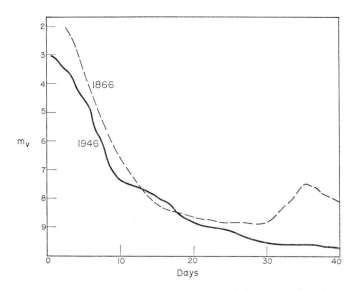

Fig. 60. Light curves of T Coronae Borealis in 1866 and 1946.

nova is assumed to be of spectral type M 2 and photographic magnitude 10.2. Outstanding features of the spectrum at this stage are the great strength of the He II emission line at 4686 Å, particularly during the early stage of the outburst, and the rapid narrowing of emission bands, together with the presence of coronal lines of [Fe X] and [Fe XIV].

Spectra taken at Merate by Gratton and Kruger (1946) in February 1946 with a slit spectrograph and prismatic camera show displaced emission and absorption components of the lines of H I, He I, and of C, N and O in various stages of ionization. Similar Doppler shifted lines were also found associated with the emission line of [O II]. The deduced colour temperature is 9000K. Almost identical spectral characteristics were found by Herbig and Neubauer (1946) and by Boroncho and Malville (1968) who also observed the coronal lines of [A X], [Fe X] and [Fe XIV].

Some years before the 1946 outburst, T Coronae Borealis was recognized as a spectroscopic binary. The spectrum, at minimum brightness, is found to be composite, consisting of a normal gM 3 type spectrum which is superimposed upon a blue continuum containing the emission lines of H I, He I, He II, O III, Ca II, [O III], [Ne III]. The width of these emission lines is fairly large ($\delta V = \sim 300$ km/sec) and a small velocity range ($K_1 = 33.5$ km/sec).

The general characteristics of this system have been given by various authors, including Sanford (1949) and Kraft (1958). The minimum masses for the red and blue components are 2.9 and 2.1 $M_\odot$ respectively. The radius of the red giant is $\sim 9.0 \times 10^{12}$ cm; orbital inclination $i \sim 65°$; $v \sin i = 1.42 \times 10^7$ cm sec^{-1} and the period is 230^d with a semi-amplitude in velocity of $K_2 = 21$ km/sec.

The small variations in brightness during minimum light are due to irregular variations in the red giant star (amplitude $\sim 0^m.5$).

RS Ophiuchi

The light curves of the four individual maxima of this star are all very similar to that

167

The Nebular Variables

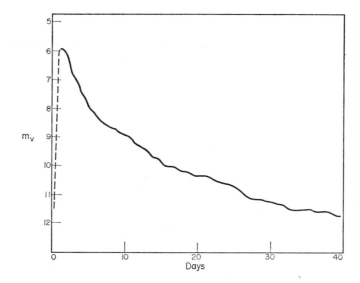

FIG. 61. Light curve of RS Ophiuchi following the outburst of October 1967. (After Mayall.)

shown in Fig. 61. Like T Coronae Borealis, RS Ophiuchi has a very fast variation in brightness during a typical outburst, the corresponding parameters defining the rate of decline being:

$$T_2 = 4^d, \quad T_4 = 15^d.$$

Extensive spectroscopic observations of this star have been made by Barbon *et al.* (1969) covering the minimum of 1959–67 and the most recent outburst in October 1967. Throughout the period of minimum light, the spectra of this star show a gradual diminution in the degree of excitation together with a progressive fading of the blue–violet continuum and a corresponding strengthening of the red region. The Balmer series from Hα to Hδ are present in emission as well as He I, while the forbidden lines of [O III] and [N II] are both very faint. The coronal lines of [Fe X] and [Fe XIV] are completely absent.

Following the explosion of 1967, the spectrum changed dramatically. Wide emission bands appeared on a strong continuum, these being bordered on the shortward side by two systems of broad absorptions (mean radial velocities of −2700 and −3900 km/sec). A very narrow absorption line was present near the centre of the emission bands, bordered on the red edge by a sharp emission. These sharp emission components had a radial velocity of ∼ −40 km/sec which is the same as the radial velocity of the star itself. As Barbon *et al.* have pointed out, it appears virtually certain that these narrow emission components originate in a stationary envelope surrounding the nova since they were also observed during the outburst of 1958.

Coincident with the decline in brightness, the absorption systems faded, eventually disappearing altogether. At the same time, the He I line increased in intensity to become the strongest line in the spectrum apart from Hα and Hβ. Lines of highly ionized atoms gradually appeared as the He I line faded and approximately 40 days after maximum, [Fe X] was very prominent.

One-hundred days after maximum, the spectra showed the very high excitation of the

168

nebular envelope ejected by the star with the most intense lines being those due to [Fe X], [A X], [Fe XIV], [O III], [Ni XIII] and [Ni XII].

Numerous spectra taken during the periods of quiescence between outbursts have shown, from the composite nature of these spectra, that the giant component in this system has a spectral type approximating to gM 2, i.e. very like that of T Coronae Borealis.

In both of these stars, as in all of the other recurrent novae, the source of the very high degree of ionization is still not known. There are several possibilities. Photoionization or the ejection of material from some region deep in the interior of the star during the outburst could account for it. On the other hand, it has been suggested that collision of the outward moving gases with the stationary envelope surrounding the star may be the cause.

IC 2220

Here we may mention the peculiar butterfly-shaped reflection nebula IC 2220 which surrounds the bright variable giant star HD 65750 = HR 3126. This star has a spectral type of M 3 III and the system has recently been examined by Dachs and Isserstedt (1973).

Although not a true symbiotic variable since there is no evidence of a companion star, nevertheless this object displays some very interesting properties. Filaments of the nebula closely resemble the structure of a magnetic dipole field centered upon the red giant.

From UBV photometry it can be demonstrated that the (B–V) and (U–B) colours of the nebula are distinctly bluer than those of the illuminating star. An extremely large loop is clearly visible in IC 2220 which Dachs and Isserstedt have tentatively interpreted as a giant prominence originating in the central star. It appears very probable that the star itself may be in the pre-main sequence stages of its evolution.

References

ALLER, L. H. (1954) *Astrophysics*, New York.
ALTENHOFF, W. J. and WENDKER, H. J. (1973) *Nature* **241**, 37.
BARBON, R., MAMMANO, A. and ROSINO, L. (1969) *Non-Periodic Phenomena in Variable Stars*, p. 257, Reidel, Dordrecht.
BELYAKINA, T. S. (1965) *Izv. Krym. astrofiz. Obs.* **33**, 226.
BELYAKINA, T. S. (1967) *Ibid.* **38**, 171.
BELYAKINA, T. S. (1969) *Ibid.* **40**, 39.
BELYAKINA, T. S. (1970a) *Astrofizika* No. 1, 49.
BELYAKINA, T. S. (1970b) *Izv. Krym. astrofiz. Obs.* **41–42**, 275.
BIDELMAN, W. P. (1954) *Astrophys. J. Suppl.* 1, No. 7.
BIDELMAN, W. P. and STEPHENSON, CH. B. (1956) *Publ. astr. Soc. Pacif.* **68**, 152.
BLOCH, M. (1952) *Ann. Astrophys.*
BLOCH, M. (1969) *Mem. Soc. Roy. Sci. Liége* **17**, 363.
BLOCH, M., JOUSTEN, N. and SWINGS, P. (1969) *Bull. Soc. Roy. Sci. Liége*, No. 5, 245.
BORONCHO, D. R. and MALVILLE, J. M. (1968) *Publ. astr. Soc. Pacif.* **80**, 177.
BOYARCHUK, A. A. (1966) *Astr. Zu.* **43**, 976.
BOYARCHUK, A. A. (1967a) *Astrofizika* 3, 203.
BOYARCHUK, A. A. (1967b) *Astr. Zu.* **44**, 12.
BOYARCHUK, A. A. (1968a) *Izv. Krym. astrofiz. Obs.* 39.
BOYARCHUK, A. A. (1968b) *Astrofizika* 4, 289.
BOYARCHUK, A. A. (1969) *Non-Periodic Phenomena in Variable Stars*, p. 395, Reidel, Dordrecht.
BOYARCHUK, A. A. (1970) *Izv. Krym. astrofiz. Obs.* **41–42**, 264.
CAPUTO, F. (1971) *Publ. astr. Soc. Pacif.* **83**, 62.
CESTER, B. (1969) *Astrophys. and Space Sci.* 3, 198.
DACHS, J. and ISSERSTEDT, J. (1973) *Astron. and Astrophys.* **23**, 241.
ELVEY, C. T. (1941) *Astrophys. J.* **94**, 140.
FARAGGIANA, R. (1969a) *Astrophys. and Space Sci.* 3, 205.
FARAGGIANA, R. and HACK, H. (1969b) *Mem. Soc. Roy. Sci. Liége* **17**, 317.

The Nebular Variables

FitzGerald, M. P. and Houk, N. (1970) *Astrophys. J.* **159**, 963.
Gauzit, J. (1955) *Ann. Astrophys.* **18**, 354.
Gratton, L. and Kruger, E. C. (1946) *Ric. Sci. Ricostruz* **16**, 299.
Greenstein, N. K. (1937) *Harvard Obs. Bull.* No. 906.
Heinze, K. G. (1952) *Astrophys. J.* **115**, 133.
Herbig, G. H. and Neubauer, F. J. (1946) *Publ. astr. Soc. Pacif.* **58**, 196.
Herbig, G. H. (1950) *Ibid.* **62**, 211.
Herbig, G. H. (1960) *Astrophys. J.* **131**, 632.
Herbig, G. H. (1969) *Proc. Nat. Acad. Sci. USA* **63**, 1045.
Himpel, K. (1940) *A.N.* **270**, 184.
Himpel, K. (1942) *Beob. Kirk.* **24**, 53.
Hodge, P. W. and Wright, F. W. (1970) *Publ. astr. Soc. Pacif.* **82**, 135.
Hogg, F. S. (1934) *P.A.A.S.* **8**, 14.
Hoffleit, D. (1968) *Irish astr. J.* **8**, 149.
Hutchings, J. D. and Redman, R. O. (1972) *Publ. astr. Soc. Pacif.* **84**, 240.
Ivanova, N. L. (1960) *Soobsc. Byurak. Obs.* **28**, 17.
Jacchia, L. (1941) *Harvard Obs. Bull.* No. 912.
Knacke, R. F. (1962) *Astrophys. Lett.* **11**, 201.
Kraft, R. P. (1958) *Astrophys. J.* **127**, 625.
Kukarkin, B. V. and Parenago, P. P. (1948) *General Catalogue of Variable Stars*, First Ed., Moscow.
Lindsay, E. M. (1932) *Harvard Obs. Bull.* No. 888, 22.
Lundmark, K. (1921) *A.N.* **113**, 94.
Mammano, A. and Rosino, L. (1966) *Mem. Soc. astr. Ital.* **37**, 493.
Mammano, A. and Rosino, L. (1968) *Ibid.* **39**, 471.
Mammano, A. and Rosino, L. (1969) *Mem. Soc. Roy. Sci. Liége* **17**, 369.
Mayall, M. (1964) *Q. Bull. A.A.V.S.O.*
McCuskey, S. W. (1965) *Ci. c. IAU* 1916, 1917.
Mendel, D. (1946) *Physica* **12**, 768.
Merrill, P. W. (1950) *Astrophys. J.* **111**, 484.
Merrill, P. W. and Burwell, C. G. (1950) *Ibid.* **112**, 72.
Merrill, P. W. (1958) *Etoiles à raies d'emission*, p. 436, Univ. of Liége.
Nguyen-n-Doan (1970) *Astron. and Astrophys.* **8**, 307.
Payne-Gaposchkin, C. (1946) *Astrophys. J.* **104**, 362.
Payne-Gaposchkin, C. (1950) *Ibid.* **115**, 411.
Payne-Gaposchkin, C. (1957) *The Galactic Novae*, North Holland Publ. Co., Amsterdam.
Petit, E., Sanford, R. F. and McLaughlin, D. B. (1946) *Publ. astr. Soc. Pacif.* **58**, 153.
Philip, A. G. D. (1969) *Ibid.* **81**, 248.
Purton, C. R., Feldman, P. A. and Marsh, K. A. (1973) *Nature Phys. Sci.* **245**, 5.
Romano, G. (1960) *Publ. Osser. astr. Padova*, 119.
Sahade, G. (1949) *Astrophys. J.* **109**, 541.
Sahade, G. (1960) *Stellar Atmospheres*, p. 494, Univ. of Chicago, Chicago.
Sahade, G. (1965) *3rd Colloquium on Variable Stars*, p. 140, Bamberg.
Sandig, H. U. (1950) *A.N.* **278**, 187.
Sanford, R. F. (1944) *Publ. astr. Soc. Pacif.* **56**, 112.
Sanford, R. F. (1949) *Astrophys. J.* **109**, 81.
Shao, C. Y. (1971) *Bull. Am. astr. Soc.* **3**, 15.
Sieder, Th. (1956) *Mitt. veränderl. Sterne*, 238.
Sobolev, V. V. (1945) *Moving Envelopes of Stars*, Moscow.
Stiemon, F. M. (1973) *Bull. Am. astr. Soc.* **5**, 17.
Swings, P. and Struve, O. (1941a) *Astrophys. J.* **93**, 356.
Swings, P. and Struve, O. (1941b) *Ibid.* **94**, 291.
Swings, P. and Struve, O. (1943) *Ibid.* **97**, 194.
Swings, J. P. and Allen, D. A. (1972) *Publ. astr. Soc. Pacif.* **84**, 523.
Swope, H. H. (1940) *Harvard Obs. Ann.* **109**, No. 1.
Tcheng Mao Lin and Bloch, M. (1952) *Ann. Astrophys.* **15**, 104.
Tcheng Mao Lin and Bloch, M. (1954) *Ibid.* **17**, 6.
Tift, W. G. and Greenstein, J. L. (1958) *Astrophys. J.* **127**, 160.
Webster, L. B. (1966) *Publ. astr. Soc. Pacif.* **78**, 136.
Wenzel, W. (1956) *Mitt. veränderl. Sterne* 227.
Wilson, R. E. (1950) *Publ. astr. Soc. Pacif.* **62**, 14.
Woolf, N. J., Stein, W. A. and Strittmatter, P. A. (1970) *Astron. and Astrophys.* **9**, 252.

CHAPTER 24

Peculiar nebular objects

HERE we shall discuss, in detail, three peculiar objects that are involved in nebulosity. Two of these stars, η Carinae and FG Sagittae, are probably related to the bright blue irregular variables discovered by Hubble and Sandage (1953) in the galaxies M31 and M33. These Hubble–Sandage variables are F-type supergiants located in a region of the Hertzsprung–Russell diagram where pulsational instability has been shown to be a dominant factor by such workers as Ledoux (1941) and Schwartzchild and Härm (1959).

A search for similar irregular variables in other galaxies has resulted in the discovery of eight more by Tammann and Sandage (1968) in NGC 2403, while it appears fairly certain from the observations of Martini (1969) that S Doradus in the Lesser Magellanic Cloud is also a member of this group.

If these variables are main sequence stars with masses $\sim 100\ M_\odot$ as suggested by Strohmeier (1972), we would expect them to possess vibrational instability since not only do they have large radii and turbulent atmospheres, but the life-time of this evolutionary stage must be extremely short.

Eta Carinae

This star was first systematically observed by Halley in 1677 when it apparently varied in an irregular manner between second and fourth magnitude. Before its observation by Halley, we knew very little of the star apart from the fact that it is not recorded in the catalogue produced by Ptolemy about 140 A.D. This suggests, although it does not prove, of course, that the star was fainter than fourth magnitude at that time.

Over the past two centuries, this object has undergone a spectacular series of irregular and unpredictable brightenings. These have been summarized by various observers including de Vancouleurs and Eggen (1952), O'Connell (1956) and Gratton (1963). Maxima of the star have been recorded in 1750, 1843, 1889, 1938 and 1950. The maximum of 1843, when the star reached $-0^m.8$, probably began as early as 1810 (Fig. 62).

The absolute visual magnitude may, of course, be estimated once the distance of η Carinae is known. This can be determined in several ways and all of them are found to be in fairly good agreement with each other.

(a) Measurement of the intensity of the interstellar lines of Ca II in the spectrum provides a good estimate of the distance although here we must make certain, reasonably valid, assumptions concerning the density of this interstellar material lying in this particular direction.

(b) A second method which can be used is to measure the spectroscopic parallaxes of the nearby stars and

The Nebular Variables

(c) The distance may be computed from the absorption features in the spectrum of this star. Thackeray (1953) has shown that gas is being continually ejected from the star and the main absorption features arise in an expanding circumstellar shell which has a velocity of expansion of ~475 km/sec.

These various estimates place η Carinae between 1.5 and 3.0 kpc away as determined by Gratton (1967) and Feinstein (1963). This implies a total luminosity of 10^{40} erg/sec. In other words, this star is, at present, 2.5×10^6 more luminous than the Sun. It is, perhaps, significant that very few other galactic objects have shown comparable luminosities for such a long period of time.

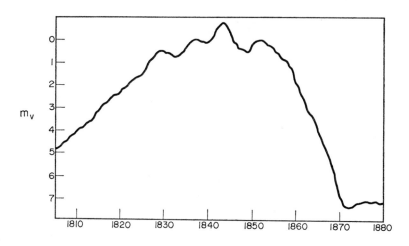

FIG. 62. Light curve of Eta Carinae.

The absolute visual magnitude of η Carinae has been estimated by Thackeray (1953) as -14^m at its peak brilliance in 1843. For several years following this event it remained around -13^m, beginning to fade only about 1860 to something like -6^m by 1870. The visual estimates of this star made prior to 1830, and some of the earlier ones made by Halley, are naturally somewhat uncertain but they, too, indicate an absolute visual magnitude of between -9^m and -11^m during the period from 1677 to 1830.

VISUAL APPEARANCE

The present visual appearance of η Carinae is of a small stellar nucleus of sixth magnitude which is surrounded by a nebulous halo in which may be seen small condensations only a few seconds of arc away from the nucleus. The very complex structure of this nebulosity has been discussed by Thackeray (1949), Gaviola (1950), Ringuelet (1958) and Gehrz and Ney (1972).

Various motions within this small reddish nebula, associated with the small condensations, have been analyzed by Ringuelet and from their velocities of expansion she has been able to show that their times of emission coincide very closely with the observed dates of light maxima. Several of the condensations were ejected at times coinciding with the maxima of 1750, 1843 and 1889 and in 1950 a new condensation was observed although this lay too close to the nucleus for accurate measurement. It should also be noted that there is clearly

172

the possibility that η Carinae may be physically connected with the nearby bright emission nebula NGC 3372.

MASS OF η CARINAE

From its distance of approximately 2.0 kpc it is quite clear that a tremendous amount of energy has been radiated from this object over the past three centuries and we are obviously dealing with a very massive star. Unfortunately, it is not an easy matter to make an accurate estimate of its mass. An approximation can be arrived at from the mass-luminosity relation but there are difficulties associated with this method since it is generally agreed that η Carinae does not lie even close to the main sequence. However, if we make the assumption that the luminosity of the star has not altered appreciably during its past history, and bearing in mind that the luminosity of a very massive star is insensitive to surface temperature, we can estimate a mass-luminosity relation as follows from the work of Stothers (1966).

$$M_* \approx 115 \left(\frac{D}{2 \text{ kpc}}\right)^{1.5} \cdot M_\odot \tag{1}$$

where D is the distance of η Carinae.

Davidson (1971) has suggested the following parameters for the central star:

Effective temperature: $T_* = 29{,}300°\text{K}$

Radius: $R_* = (4.4 \times 10^{12} \text{ cm}) \left(\frac{D}{2 \text{ kpc}}\right)$

Luminosity: $L_* = (10^{40} \text{ erg s}^{-1}) \left(\frac{D}{2 \text{ kpc}}\right)^2$

Ionizing photons: $S_i = (4 \times 10^{49} \text{ s}^{-1}) \left(\frac{D}{2 \text{ kpc}}\right)^2$

where, again, D is the distance of the star. While these are, admittedly, approximate values, they do show that the star must be extremely massive.

THE SPECTRUM OF η CARINAE

While the visible spectrum of this star undoubtedly provides us with a great deal of information, it is unfortunate that the variations in its spectroscopic characteristics have been observed for a much shorter period than its light changes. We have, for example, no knowledge at all of the changes in the spectrum that must have taken place during the very important period from 1810 to 1860 which covers the time when it was at its most brilliant.

A comprehensive study of the visible spectrum of η Carinae has been made by several observers including Aller and Dunham (1966), Rodgers and Searle (1967), Neugebauer and Westphal (1968) and Westphal and Neugebauer (1969). Numerous emission lines are present, particularly those of the Balmer series and [Fe II], together with Ti I although the latter generally appears in absorption rather than emission. The emission lines are normally quite sharp although broad emission features superimposed upon the continuum are also present, these yielding an expansion velocity of ~ 400 km/sec. Several of the broader emission bands possess absorption components on their violet edges.

Most of the observed energy flux lies in the far infrared and the infrared spectrum has been examined by Thackeray (1953) who has succeeded in identifying the sharp, strong

173

The Nebular Variables

lines of the Paschen series together with those due to Ne I, Ti II and A III. The lines of other metals have also been found in this region of the spectrum as well as several very intense lines which, so far, have resisted identification.

Infrared spectral scans of this star in the wavelength range from 8 to 13μ together with broad-band photometric measurements have been made recently by Robinson *et al.* (1973) which show that the spectrum in this region exhibits a feature characteristic of thermal emission from silicate grains.

The position angle of the electric vector of polarization and also the degree of polarization have been measured by Visvanathan (1967) and compared with those of nearby stars. Both differ appreciably for η Carinae, the degree of polarization being almost double that of other stars in the neighbourhood. One further striking feature of this object is that it is a γ-ray source comparable with the Crab Nebula. McCray (1967) has calculated the energy to be somewhat greater than 5×10^{22} eV. There is also a strong radio flux at 1 cm but, curiously, no detectable emission at either 11 or 20 cm.

The spectral type of the central star itself, as distinct from that of the surrounding nebula, is cF5.

THE NATURE OF η CARINAE

Several theories have been put forward to explain the nature of this peculiar object. These may be summarized as follows:

(a) As suggested by Gratton (1967), this may be a very young star which is still evolving towards the main sequence. If it is still condensing from the surrounding protostellar cloud we can theorize that, in addition to vibrational instability, there will be violent convective and chromospheric activity taking place.

(b) It may be a very massive star that has reached and is now evolving away from the main sequence. Because of its mass, such a star is likely to be pulsationally unstable as shown by Burbidge (1962) and Talbot (1971). Burbidge has also shown that it is theoretically possible for η Carinae to evolve by mass loss, either by nuclear reactions or ejection into the surrounding envelope, without undergoing the catastrophic explosion of a supernova. The classical contraction time of a star of $\sim 100 \ M_\odot$ is 1.6–2.0×10^4 years, whereas the total life span for a star of this mass and absolute magnitude lies between 2.0 and 7.0×10^6 years.

(c) It is possible that this object may be a non-thermal or non-stellar object, perhaps the remains of a recent massive supernova, as suggested by Zwicky (1965) and Ostriker and Gunn (1971).

The presence of a circumstellar shell of dust around η Carinae is known from the reddening of this star ($E_{B-V} \approx 1^m.1$) which is appreciably greater than the interstellar figure of $E_{B-V} \approx 0^m.44$ observed by Feinstein (1963, 1969). Davidson (1971) has confirmed the earlier suggestion by Pagel (1969) that the stellar continuum and the far infrared spectrum contain the same energy flux. It would therefore seem that the infrared represents thermal re-radiation by this dust of much of the visible and ultraviolet radiation of the central star.

From the general characteristics of the infrared spectrum, which covers quite a wide range of frequencies, it seems that the temperatures of the surrounding dust grains also cover a wide range from ~ 150 to $\sim 1500°$K.

The polarization as measured by Visvanathan (1967) is also consistent with scattering by dust grains. The fact that this polarization is found to be almost independent of wavelength, however, implies that the grain sizes are quite small as indicated by Hanner (1971).

174

From the above, therefore, η Carinae appears to be an extremely massive star which is surrounded by an envelope of dust and gas which is clearly expanding.

If it is a very young object evolving towards the main sequence, then the envelope may have been produced by the outer layers of the initial protostar which are being expelled from the star by radiation pressure acting upon the dust grains as expressed by Davidson and Harwit (1967), and Davidson (1971). It is difficult, however, to apply the necessary time scale to such a model.

Another alternative is suggestion (b) given above, namely that here we have a star evolving away from the main sequence. The behaviour of stars with masses of the order of 100 $M_\odot$ has been examined by Appenzeller (1970), Simon and Stothers (1970), Ziebarth (1970) and Talbot (1971). Pulsational instability will occur while such stars are on the main sequence leading to ejection of mass, but this instability may cease once hydrogen is depleted at the stellar core.

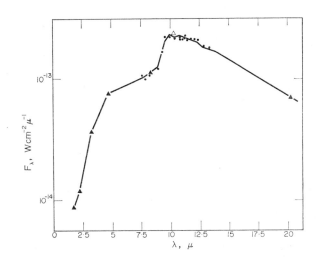

FIG. 63. Infrared energy distribution of Eta Carinae. Filled triangles represent the flux found from broad-band photometric observations and solid dots those from spectrophotometric data obtained in March 1972. The open triangle is the 10.2 μ flux corrected for the redness of this object. (After Robinson *et al.*)

As Davidson (1971) has also shown, evolutionary expansion leads to a lowering of the surface temperature and radiation pressure is then likely to bring about some mass ejection.

It is possible, as described by Appenzeller (1970), that η Carinae may still be on the main sequence although appearing to lie above it. If there is pulsation, there may also be continuous ejection of material, resulting in the formation of an opaque envelope around the star. If this is so, then the measured temperature and radius of η Carinae will be those of the surrounding envelope and not of the star itself. Consequently, while the star lies on the main sequence, the object will seem to lie above it.

The observations made in the 8–13μ infrared region by Robinson *et al.* (1973) have led to a re-interpretation of the original dust shell model for this star in terms of a more complicated picture of the dust envelope. The observed infrared spectral energy distribution of this star is shown in Fig. 63.

The Nebular Variables

No emission lines, for example the [S IV] line at 10.53μ and that of [Ne II] at 12.8μ, were found although Gillett and Stein (1969) and Rank *et al.* (1970) have shown these to be present in the spectra of certain planetary nebulae.

The spectral energy distribution in the $8–13\mu$ region, as determined by Robinson *et al.* (1973), is of particular importance. As shown in Fig. 64 there is not a smooth continuum but rather a sharp rise between 8 and 9μ, followed by a gradual decline.

The close similarity between the wavelength dependence of the excess with the emissivity of enstatite (silicate grains) as determined by Gaustad (1963) and the smoothed μ Cephei excess measured by Woolf and Ney (1969) provides evidence that large numbers of silicate-like grains are present in the shell surrounding η Carinae.

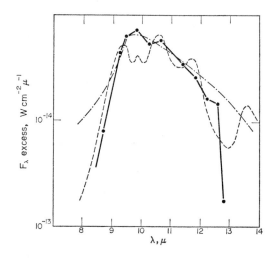

FIG. 64. The excess emission from Eta Carinae over a 230°K black body fitted to the 8 and 13μ points, shown as solid dots. The emissivity of enstatite (———) and the excess emission from μ Cephei (–·–·–·–) are also shown. (After Robinson *et al.*)

From the observed spectral energy distribution Robinson *et al.* have shown that it is no possible to obtain a satisfactory fit from either a single temperature blackbody or the more advanced models of Krishna Swamy (1971) and Neugebauer *et al.* (1971).

A double-shell model provides a much better fit and such a model is clearly consistent with the past history of this object and the optical structure of the surrounding nebulosity. On the basis of this model, it appears that most of the infrared flux comes from the inner shell between 2.0 and 3.6μ, whereas the outer shell provides the major proportion of the infrared flux between 4.8 and 10.2μ.

X-RAYS FROM η CARINAE

Recently, a soft X-ray flux has been detected from this object by Hill *et al.* (1972) between 0.2 and 3 keV. A thermal exponential with kT ~ 250 eV approximately fits this spectrum if it is assumed that a low energy interstellar cut-off may be invoked. This corresponds to a temperature of the order of 3×10^{6}°K. Since both X-ray and ultraviolet emission lines are especially important at this very high temperature, it is not strictly satisfactory to adopt an exponential spectrum here, but as shown by Davidson and Ostriker (1972) most of the line radiation fits the spectrum and, if the emitting gas is thermal, its temperature must be in the range of several million degrees.

Taking the distance of this object as 2 kpc and making due allowance for interstellar absorption, we find that the total X-ray luminosity of η Carinae is $\sim 3 \times 10^{36}$ erg sec^{-1}.

Whatever the nature of the central object, it is clearly surrounded by a shell of material which has been ejected during the outbursts from 1750 onward. Radial motions in this material are some hundreds of km/sec. The expanding shell is no longer spherical in shape although it still possesses a well-defined outer boundary with major and minor axes of 3×10^{12} and 1.5×10^{12} km respectively. Inside this region there exists gas with an effective temperature of $T_e \sim 7500°$K and an electron density of $n_e = 4 \times 10^6$ cm^{-3}, this gas probably being photoionized according to Davidson (1971).

CONSTRAINTS ON MODELS OF η CARINAE

Davidson and Ostriker (1972) have examined the probable situation present in the neighbourhood of this object producing the observed soft X-ray flux with a view to deciding between the choice of models; a massive, largely intact star as against a supernova remnant.

Following Davidson and Ostriker, it is assumed that a shock front is forced into the surrounding medium by the expanding gas, there being a zone between the expanding shell and the shock front which is occupied by the gas that is swept up, compressed and heated. Conservation conditions across this front lead to the following equations;

$$w^2 = 3v_1{}^2/16 \tag{2}$$

$$v_2 = v_1/4 \tag{3}$$

$$n_2 = 4n_1 \tag{4}$$

where w is the isothermal velocity of sound in the hot gas, v_1 is the shock front velocity into the ambient, relatively cool medium, v_2 is the velocity of the adiabatically compressed hot gas relative to the shock front, and n_1, n_2 are the undisturbed and compressed densities in the swept-up gas.

From observation, v_1 is found to be ~ 500 km/sec and therefore $w \sim 220$ km/sec which corresponds to a temperature of $3.4 \times 10^{6°}$K as suggested by the X-ray flux. In addition, the density of the hot gas is estimated as $n_2 = 8000$ cm^{-3} and the cooling time as ~ 80 years. From the latter it is concluded that much of the gas that has been swept up following the outburst of 1843 is still comparatively hot.

The fact that the X-ray luminosity is $\sim 3 \times 10^{36}$ erg sec^{-1} implies that the hot gas must occupy a volume of about 1.6×10^{36} km^3. Such dimensions are not impossibly high, implying a shell with a mean radius of 10^{12} km and a thickness of 1.3×10^{11} km which would lie just on the periphery of the observed shell of material ejected from the star. Certainly the mass of the hot gas ($\sim 0.01\ M_\odot$) is much smaller than that of the photoionized gas at 7500°K but here it is not possible to estimate the amount of invisible, non-ionized material that may be present in the expanding shell. From direct photographic observation of the region close to η Carinae, it would seem that the dense gas at 7500°K is in the form of small filaments and condensations rather than a simple shell. Nevertheless, it is clear that the expanding gas must be moving into a surrounding medium of high density given by $n_1 \approx n_2/4$ (from equation 4) ≈ 2000 cm^{-3}.

How do these observations affect the choice between models? First, Davidson and Ostriker have pointed out that the very steep slope of the X-ray spectrum suggests that it is thermal in origin and any non-thermal origin is correspondingly quite small. According

The Nebular Variables

to the supernova remnant model derived by Ostriker and Gunn (1971), a flatter spectrum at high energies is required since the non-thermal X-ray luminosity would be greater than 10^{38} erg sec^{-1}. In addition, the observed visual emission lines indicate that the 7500K gas is compressed to high densities and we may invoke the shock front pressure, representing the deceleration brought about by the sweeping up of material, to keep this gas in a highly compressed state.

On the assumption that η Carinae is an extremely massive star ($\sim 60\,M_\odot$), it is difficult to see how such an object can be fitted into some of the current theories of stellar formation, for example, that due to Larson and Starrfield (1971). The possibility exists that η Carinae may be related in some way to the far infrared sources near the galactic centre postulated by Hoffmann et al. (1971).

LkHα-101

At this point it is of interest to mention the highly reddened star LkHα-101 which lies within the reflection nebula NGC 1579 and which was first noticed to show Hα in emission by Herbig (1956). Little attention was paid to this object until the discovery by Cohen and Dewhirst (1970) and Cohen and Woolf (1971) that it has a very intense infrared flux.

The similarity between LkHα-1010 and η Carinae in the photographic infrared was shown by Herbig (1971) who found many prominent emission lines in this region including the Paschen series, He I, N I, O I, Fe II, [Fe II], [O II], [Cr II] and other unidentified lines, which are also present in the infrared spectrum of η Carinae.

Allen (1973), however, has shown that in the visible region, the spectrum of LkHα-101 is dominated by the continuum and, apart from emission lines of Hα and Hβ, all others are very weak. Eta Carinae, on the other hand, has a visible spectrum dominated by emission lines.

The spectral type as determined from the absorption features is F8: e II, while the luminosity has been estimated by Allen (1973) as approximately 4×10^3 L$_\odot$. This figure is typical of a B type star and far lower than that of η Carinae. All of the evidence, therefore, is against the idea that LkHα-101 is a northern counterpart of Carinae.

FG Sagittae

Like η Carinae, this star shows a P Cygni type of spectrum indicative of the outflow of material into a surrounding envelope.

In 1894 this star had a photographic magnitude of 13^m.6, but since that time it has brightened extremely slowly until it is now 9^m.4. Superimposed upon this slow brightening are small and irregular fluctuations (Fig. 65).

Several spectrograms have been taken of this object during the period from 1955 to 1967 and these have been described by Herbig and Boyarchuk (1968) who have shown that the basic spectral type has changed in step with the light variations, progressing from B4 I, through B9 Ia and A3 Ia, to its present type of A5 Ia. At the same time, the bolometric magnitude has increased from $-3^m.6$ to $-4^m.3$. The spectra all indicate a high luminosity for this star and show typically strong P Cygni type emission fringes on the Balmer lines.

It has also been possible to estimate the radius of the photosphere around this object and, again, we find a gradual increase during the same period of from 10 $R_\odot$ in 1955 to 26 $R_\odot$ in 1965.

178

Almost certainly we have here the slow ejection of a shell by a central star which is possibly a supergiant of Population I. The star itself is invisible. While it is not possible at the present time to measure the velocity of expansion of the surrounding nebulosity accurately, it would appear to be something of the order of 70–80 km/sec, corresponding to a period of about 3000 years since its formation.

Long exposure photographs show that the object is centered upon a normal planetary nebula with a radius of 18″ but this nebula was clearly in existence long before the observed brightening of 1894. The most plausible situation in the case of FG Sagittae is that some 3000 years ago this planetary nebula was formed and a second one is now in the process of formation.

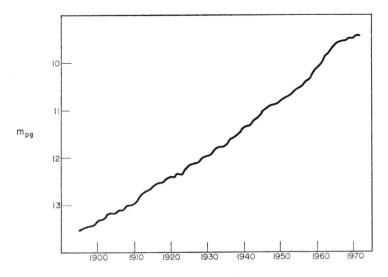

FIG. 65. Light curve of FG Sagittae.

FU Orionis

As we have seen earlier, it is extremely likely that the T Tauri variables are stars which are gravitationally bound to their parent nebulae. We must here consider the variable FU Orionis which represents a very real, and significant, association of a star and interstellar material.

In 1936 this star rose fairly abruptly from an apparent magnitude of $16^m.2$ to $10^m.1$. The rate of brightening then slowed somewhat with the star reaching $9^m.7$ by late 1937. It is quite probable that the peculiar light variations which were noticed prior to 1938 were caused by large flares which, in themselves, could bring about the necessary spallation reactions.

Photographs show that FU Orionis lies within the heavy obscuration nebula Barnard 35 and there is little doubt that the two are physically connected.

THE SPECTRUM OF FU ORIONIS

The visible spectrum of this star shows several peculiarities, resembling in many ways that of F-type stars of high luminosity (F2 0 I–II), but the Balmer lines are abnormally intense in absorption. Undoubtedly, this is due to the superposition of lines from two different sources; that of the star and of the shell which, as we shall see below, have an overlap of 80 km/sec.

In the shell spectrum we find strong lines due to neutral metals such as Na I, Al I and Cr I, but these are extremely weak in the stellar spectrum. The neutral lithium line of Li I at 6707 Å is also present in considerable strength and it is possible that this element may be formed by spallation of high energy protons and heavier elements as suggested by Fowler *et al.* (1962).

The radial velocities that have been measured vary from line to line, ranging from -50 km/sec for the shell to $+30$ km/sec for the star.

THE NATURE OF FU ORIONIS

The probable nature of this star has been reviewed by Herbig (1966). It is clearly a large and luminous object with the following approximate parameters:

Absolute visual magnitude (maximum): -1^m to -2^m.
Radius: 20–$25\ R_\odot$.
Luminosity (circumstellar star): $\sim 1000\ L_\odot$.

Judging from its past history, it is usually considered that this object was initially a very large protostar with a radius of $\sim 100\ R_\odot$ which then collapsed to its present size.

An alternative theory of the evolution of FU Orionis has been put forward by Larson (1972) based upon the non-homologous collapse of a protostellar cloud.

Although the spectrum of this star is clearly abnormal it does indicate that the effective temperature is of the order of $7 \times 10^{3}\,^\circ$K. This is similar to that of another variable which, in many respects, is very like FU Orionis, namely LkHα-190 (V1057 Cygni). Like the former star, V1057 Cygni has been observed to brighten by $\sim 5^m$ within a year and the effective temperature (calculated from its spectrum) is about $10^{4}\,^\circ$K.

In both bolometric luminosity and temperature of the central stellar object, these two stars show a close resemblance to R Monocerotis. The major difference is that in R Monocerotis, most of the stellar radiation is absorbed by the dense circumstellar shell and reradiated in the infrared. This does not appear to be the case for either FU Orionis or V1057 Cygni.

Cohen and Woolf (1971) have pointed out that the changes noted in the spectrum of V1057 Cygni during the rapid increase in brightness are not inconsistent with the comparatively abrupt disappearance of a circumstellar shell.

Larson (1972) has suggested that an object such as FU Orionis would result from the sudden removal of the circumstellar envelope from around a star such as R Monocerotis. On the basis of the non-homologous collapse of a protostellar cloud as put forward to explain the properties of the genuine nebular variables discussed earlier, no change in the structure of the central star can lead to such a sudden increase in luminosity by a factor of 100 or more.

However, the dynamical time scale for the circumstellar shell around the central core is approximately 1 year which is comparable with the time of brightening of both FU Orionis and V1057 Cygni. Here we must examine the process whereby the circumstellar shell can be permanently dissipated. In the case of the T Tauri variables, for example, there are no rapid changes in the make up of this shell since it is being continually replenished by infall of matter from the outermost regions of the protostellar cloud. For the effects observed for FU Orionis, the accretion process must be terminated rapidly.

Such a rapid termination of this process can occur due to the action of radiation pressure on the infalling material. Radiation pressure will eventually become dominant over gravity provided the protostar is sufficiently luminous and the infalling cloud has grown optically thin. This has been shown by Larson and Starrfield (1971), according to whom the ratio of radiation pressure to gravity in an optically thin region is given by

$$\left|\frac{\text{radiation pressure}}{\text{gravity}}\right| = 7.8 \times 10^{-5} \kappa \frac{L/L_\odot}{M/M_\odot} . \tag{5}$$

Following from this equation, if a value for κ of ~ 250 cm^2 g^{-1} is adopted for the opacity of the interstellar material at visual wavelengths, the radiation pressure is dominant if

$$\frac{L/L_\odot}{M/M_\odot} > 50 \tag{6}$$

which is found to be the case for masses greater than $\sim 3\ M_\odot$.

Now the masses of FU Orionis and V1057 Cygni, as implied from their luminosities, are of the order of 5 $M_\odot$. Thus they both lie in the mass range where the infall of matter will eventually be halted and then reversed by radiation pressure. It is assumed, of course, that there is always a strong frictional coupling between the gas and dust grains.

In the outermost regions of the protostellar cloud the radiation pressure does not assume any real importance until an appreciable amount of visual radiation escapes (i.e. when the interstellar obscuration has been lifted sufficiently). Any temporary disruption of the shell, however, will halt the infall of material and permanent disappearance of the shell will then occur due to non-replenishment by infalling matter.

If we determine the point at which the radiation pressure acting upon the shell becomes of the same order as the dynamical pressure exerted by the infalling material, we find when the radiation pressure begins to affect the dynamics of the shell. For a star of 5 $M_\odot$ and a dust evaporation temperature of 2000°K, this point is reached when the infalling cloud has an optical depth of ~ 6 and a mass of $\sim 0.05\ M_\odot$. Dissipation of the shell then takes place on a time scale of approximately 1 year and there will be an accompanying brightening of the object by about 6^m which is in good agreement with the observed behaviour of FU Orionis and V1057 Cygni.

A somewhat different theory has been put forward by Pismis (1971) concerning V1057 Cygni. Here it is argued that the brightening of this object and the gradual change in spectral type from K0 to A1 may be explained by a fast readjustment of the mass distribution in the star. It can be shown that if this should occur there will be a subsequent release of potential energy in the form of visible radiation.

This readjustment of the mass is believed to have been triggered off by a significantly violent non-thermal event. Distance estimates indicate that this object is actually embedded within the North America Nebula (NGC 7000).

References

ALLEN, D. A. (1973) *Mon. Not. Roy. astr. Soc.* **161**, 1P.
ALLER, L. H. and DUNHAM, T. (1966) *Astrophys. J.* **146**, 126.
APPENZELLER, I. (1970) *Astron. and Astrophys.* **5**, 355.
BURBIDGE, G. R. (1962) *Astrophys. J.* **136**, 304.
COHEN, M. and DEWHIRST, D. W. (1970) *Nature* **228**, 1077.
COHEN, M. and WOOLF, N. J. (1971) *Astrophys. J.* **169**, 543.

The Nebular Variables

DAVIDSON, K. and HARWIT, M. (1967) *Ibid.* **148**, 443.
DAVIDSON, K. (1971) *Mon. Not. Roy. astr. Soc.* **154**, 415.
DAVIDSON, K. and OSTRIKER, J. P. (1972) *Nature Phys. Sci.* **236**, 46.
DE VANCOULEURS, G. and EGGEN, O. J. (1952) *Publ. astr. Soc. Pacif.* **64**, 185.
FEINSTEIN, A. (1963) *Publ. astr. Soc. Pacif.* **75**, 492.
FEINSTEIN, A. (1969) *Mon. Not. Roy. astr. Soc.* **143**, 273.
FOWLER, G. A., GREENSTEIN, J. L. and HOYLE, F. (1962) *Geophys. J. RAS* **6**, 148.
GAUSTAD, J. E. (1963) *Astrophys. J.* **138**, 1050.
GAVIOLA, E. (1950) *Ibid.* **111**, 408.
GILLETT, F. C. and STEIN, W. A. (1969) *Astrophys. J. Lett.* **155**, L97.
GRATTON, L. (1963) *Star Evolution*, p. 297, Academic Press, New York.
GRATTON, L. (1967) *Star Evolution*, p. 297, Academic Press, London and New York.
HANNER, M. S. (1971) *Astrophys. J.* **164**, 425.
HERBIG, G. H. (1956) *Publ. astr. Soc. Pacif.* **68**, 353.
HERBIG, G. H. (1966) *Vistas in Astronomy* **8**, 109.
HERBIG, G. H. and BOYARCHUK, A. A. (1968) *Astrophys. J.* **153**, 397.
HERBIG, G. H. (1971) *Ibid.* **169**, 537.
HILL, R. W., BIRGINYON, G., GRADER, R. J., PALMIERI, T. M., SEWARD, F. D. and STOERING, J. P. (1972) *Ibid.* **171**, 591.
HOFFMANN, W. F., FREDERICK, C. L. and EMERY, R. J. (1971) *Astrophys. J. Lett.* **164**, L23.
HUBBLE, E. and SANDAGE, A. (1953) *Astrophys. J.* **118**, 353.
KRISHNA SWAMY, K. S. (1971) *Observatory* **91**, 120.
LARSON, R. B. and STARRFIELD, S. (1971) *Astron. and Astrophys.* **13**, 190.
LARSON, R. B. (1972) *Mon. Not. Roy. astr. Soc.* **157**, 121.
LEDOUX, P. (1941) *Astrophys. J.* **94**, 537.
MARTINI, A. (1969) *Astron. Ap.* **3**, 443.
McCRAY, R. (1967) *Astrophys. J.* **147**, 544.
NEUGEBAUER, G. and WESTPHAL, J. A. (1968) *Astrophys. J. Lett.* **152**, L89.
NEUGEBAUER, G., BECKLIN, E. E. and HYLAND, A. R. (1971) *A. Rev. Astr. Astrophys.* **9**, 67.
O'CONNELL, D. J. K. (1956) *Vistas in Astronomy* **2**, 1165.
OSTRIKER, J. P. and GUNN, J. E. (1971) *Astrophys. J. Lett.* **164**, L95.
PAGEL, B. E. J. (1969) *Astrophys. Lett.* **4**, 221.
PISMIS, P. (1971) *Bol. Obs. Tonantzintla Tacubaya* **6**, 131.
RANK, D. M., HOLTZ, J. Z., GEBALLE, T. R. and TOWNES, C. H. (1970) *Astrophys. J. Lett.* **161**, L185.
RINGUELET, A. (1958) *Z. fur Astrophys.* **46**, 276.
ROBINSON, G., HYLAND, A. R. and THOMAS, J. A. (1973) *Mon. Not. Roy. astr. Soc.* **161**, 281.
RODGERS, A. W. and SEARLE, L. (1967) *Ibid.* **135**, 99.
SCHWARTZCHILD, M. and HÄRM, R. (1959) *Astrophys. J.* **129**, 637.
SIMON, N. R. and STOTHERS, R. (1970) *Astron. and Astrophys.* **6**, 183.
STOTHERS, R. (1966) *Astrophys. J.* **144**, 959.
STROHMEIER, W. (1972) *Variable Stars*, p. 33, Pergamon Press, Oxford.
TALBOT, R. T. (1971) *Astrophys. J.* **165**, 121.
TAMMANN, G. A. and SANDAGE, A. (1968) *Ibid.* **151**, 825.
THACKERAY, A. D. (1953) *Mon. Not. Roy. astr. Soc.* **113**, 211.
WESTPHAL, J. A. and NEUGEBAUER, G. (1969) *Astrophys. J. Lett.* **156**, L45.
WOOLF, N. J. and NEY, E. P. (1969) *Ibid.* **155**, L181.
VISVANATHAN, N. (1967) *Mon. Not. Roy. astr. Soc.* **135**, 275.
ZIEBARTH, K. (1970) *Astrophys. J.* **162**, 947.
ZWICKY, F. (1965) *Stellar Structure*, p. 140, Univ. of Chicago Press, Chicago.

CHAPTER 25

Infrared stars

THE LAST two decades have seen tremendous advances in the development of multifilter broad- and narrow-band photometers and bolometers specifically designed for the measurement of stellar radiation over a wide region of the infrared spectrum. Indeed, it is true to say that a far wider range of astronomical phenomena has been observed in this spectral region than was originally expected.

In addition, observations in the microwave region have revealed the lines due to several molecular species in a number of cool interstellar clouds; these transitions having been detected both in emission (often with evidence of maser amplification) and in absorption.

As far as stellar objects are concerned, some of these infrared sources are clearly very young stars (collapsing protostars) similar to the nebular variables, in which virtually none of the stellar radiation from the central core penetrates the surrounding dust cloud. Others are highly evolved stars, normally red giants, which are shrouded by dust shells of their own fabrication.

From their positions on the Hertzsprung–Russell diagram, the central sources of the latter objects are consistent with their having evolved away from the main sequence as stars with masses in the range 8–12 $M_\odot$. As pointed out by Feldman et al. (1969), many of the infrared objects are more likely to be highly evolved rather than young formations since the time scale for nuclear evolution is appreciably longer than that for gravitational collapse. We also find a long period variability at 2μ in many of the bright infrared sources ($P = 100$–200 days; amplitude $\sim 2^m$) which is very like that found in the optical region for the long period and semi-regular variables, all of which are believed to be evolved stars.

Some cool M type giant and supergiant stars possess a pronounced infrared excess in both the 10 and 20μ region over that calculated for the black body spectrum. These correspond very closely to the positions for the stretching and bending resonances of the SiO bonds and may be due to the presence of silicate grains in the extensive atmospheres of these stars. However, such irregular and long period variable stars radiate only a small part of their energy (typically less than 1 per cent of the total) in the infrared excess. Gillett et al. (1968) have reported a marked deficiency in the infrared flux at $\sim 8\mu$ in many of these objects but, as shown by Woolf and Ney (1969), this is almost certainly a spurious effect.

Pre-Main Sequence Infrared Objects

During the early phases of its evolution, the stellar core at the centre of a cloud of infalling material will be completely obscured by this enshrouding dust envelope. Consequently, none of its visible radiation manages to shine through and it is detectable only as an infrared object. A large number of such sources have been discovered by Neugebauer and

The Nebular Variables

Leighton (1965), most associated with cometary nebular and young clusters. Among the latter are such nebulae as IC 5146 and NGC 2264 which have been shown to be very young objects by Walker (1956, 1959) and the VI Cygni association which is considered by Reddish *et al.* (1967) to be an extremely young globular cluster where star formation is currently going on.

Cohen (1973) has carried out extended infrared observations on a large number of young objects in these stellar regions as well as in aggregates of emission stars near NGC 7000 and IC 5070.

The grouping in NGC 7000 contains the object LkHα-190 which according to Herbig (1958) has a T Tauri like spectrum and which recently emulated FU Orionis by brightening fairly rapidly with an amplitude of $\sim 6^m$. The optical behaviour of this object has been described by Welin (1971). The first infrared observations of LkHα-190 were made in 1971 by Cohen and Woolf (1971) and later estimates by Simon *et al.* (1972) and independently by Rieke *et al.* (1972) have shown a fairly steady diminution in the brightness at both optical and infrared wavelengths, particularly the intense features at 10 and 20μ. The suggestion by Rieke *et al.* (1972) that the decrease in the intensity of the feature at $\sim 20\mu$ is due to a cooling of dust around this object is not borne out by the observations of Cohen (1973) who has demonstrated that the dust temperature was $189 \pm 12°$K in March 1971 and $195 \pm 14°$K in October 1971, assuming that the [11.3]–[18] indices of this star may be taken as representing the colour temperatures of the dust particles and their emissivities are approximately the same at these two wavelengths.

The possibility that objects such as FU Orionis and LkHα-190 represent dust-enshrouded protostars in which the surrounding envelope is abruptly disrupted and dispersed by a combination of radiation pressure and mass ejection has already been discussed. We shall now concern ourselves with those objects which emit almost all of their energy in the infrared with very little showing in the optical region of the spectrum.

Calculations made by Larson (1972) show that the opaque phase in such objects persists for between $\sim 3.5 \times 10^5$ and 10^6 years following the formation of the stellar core in the centre of a collapsing protostellar cloud (i.e. it is comparable with the free fall time of the original cloud). Because of the relative shortness of this period, and the fact that only the brighter of these objects are likely to be detected in any survey, the number of known infrared objects is expected to be quite small. In spite of this, a fair number have been discovered, a proportion of which appear to be protostars going through these early evolutionary phases.

The Emitted Spectrum of a Protostar

Larson (1969a) has calculated the emitted spectrum of a protostar for several cases with different choices for some of the important parameters. Here we shall confine ourselves to two of these cases.

(a) A total mass of $2\ M_\odot$ for the protostellar cloud. The initial density and temperature are taken as 2.5×10^{-19} g cm^{-3} and $10°$K respectively, and the boundary conditions taken are $P = $ constant until the formation of the stellar core and the flow velocity $= 0$ (i.e. a fixed boundary) thereafter.

(b) A total mass of $5\ M_\odot$ for the protostellar cloud. The initial density and temperature being taken as 4.4×10^{-18} g cm^{-3} and $100°$K respectively. The boundary conditions are the same as for the previous case.

THEORY

Following Larson (1969b), the emitted spectrum is calculated from considerations of the radiation transfer in the infalling cloud using a simple model for the extended spherical "atmosphere" of the protostar. This model is based upon the following assumptions:

(1) The relation between radius and the product of the density (ρ) and opacity (κ_λ) follows an inverse power law of the form

$$\kappa_\lambda \rho = K_\lambda r^{-n}. \tag{1}$$

If the dust absorption properties do not vary with the radius, it can be shown (Larson, 1969a) that ρ is very closely proportional to $r^{-3/2}$ throughout virtually the whole region where radiative transfer is important.

(2) Since there is no precise information regarding the dust absorption coefficient at infrared wavelengths and high temperatures, the following simple law is assumed.

$$\kappa_\lambda = \kappa_0 \lambda^{-p} \tag{2}$$

where κ_0 and p are regarded as free parameters. The most relevant data here, on the assumption that the dust grains consist primarily of graphite and wavelengths greater than $\sim 1\mu$, is that due to Hoyle and Wickramasinghe (1962) who have shown that, under these conditions, $\kappa \propto \lambda^{-2}$.

(3) The scattering of radiation by the dust grains is ignored since at long infrared wavelengths this is negligible although this is certainly not so for visual wavelengths. Only pure absorption and thermal emission are considered.

In order to calculate the total luminosity emitted at each wavelength it is necessary to integrate the contributions from all regions of the surrounding cloud, including an attenuation factor of the form $e^{-\tau}$ for each volume element and each line of sight. This can be written in the following form;

$$L_\lambda = 16\pi^2 \int_0^\infty r^2(\tau_\lambda) G(\tau_\lambda) B_\lambda(\tau_\lambda) \, d\tau_\lambda \tag{3}$$

where

$$\tau_\lambda = \int_\tau^\infty \kappa_\lambda \rho \, dr \tag{4}$$

and

$$G(\tau_\lambda) = \frac{1}{2r} \int_0^r \frac{s\,ds}{\sqrt{(r^2 - s^2)}} \left[\exp\left\{ -\int_s^\infty \kappa_\lambda \rho \, \frac{r\,dr}{\sqrt{(r^2 - s^2)}} - \int_s^\infty \kappa_\lambda \rho \, \frac{r\,dr}{\sqrt{(r^2 - s^2)}} \right\} \right.$$
$$\left. + \exp\left\{ -\int_r^\infty \kappa_\lambda \rho \, \frac{r\,dr}{\sqrt{(r^2 - s^2)}} \right\} \right]. \tag{5}$$

In the above equations, L_λ is the luminosity emitted per unit wavelength interval, τ_λ is the radial optical depth at wavelength λ, and $B_\lambda(\tau_\lambda)$ is the black body radiation intensity function. The function $G(\tau_\lambda)$ is defined such that $G(0)=1$. Defining $G_n(\tau_\lambda)$ as the function obtained by substitution of equation (1) into equation (5), Larson (1969b) has calculated

The Nebular Variables

the function $G_{3/2}(\tau_\lambda)$ numerically which may then be used in conjunction with equation (3) to calculate the emitted spectrum L_λ, provided we know the black body function $B_\lambda(\tau_\lambda)$ or equivalently the temperature distribution $T(\tau_\lambda)$.

The simplest way to construct an approximate formula for the temperature distribution at all optical depths (Larson, 1969b) is to sum the expressions for T^4 given in equations (6) and (7) which refer to the optically thin and thick limits respectively. For the optically thin region we find that

$$T^4 = ar^{-8/(4+p)} \qquad (6)$$

where a is a constant that depends on the emitted spectrum. In the limit of large optical depths, the corresponding result is

$$T^4 = \frac{3(n + p/2 - 1)}{(n + 1)} \tau_R T_f^4 \qquad (7)$$

where

$$\tau_R = \int_r^\infty \kappa_R \rho \, dr, \quad T_f^4 = \frac{L}{16\pi\sigma r^2} .$$

The sum of equations (6) and (7) may be written as

$$T = T_0 f(\tau_R) \qquad (8)$$

where

$$T_0^4 = \left[a^{-4}\beta^{-P} \left(\frac{2}{K}\right)^4 \left(\frac{L}{16\pi\sigma}\right)^{2n-2} \right]^{\alpha/(4-p)} \qquad (9)$$

and

$$f(\tau_R)^4 = \beta[A\tau_R^{-\alpha p/(4+P)} + \tau_R]\tau_R\alpha. \qquad (10)$$

Here τ_R is related to the optical depth τ_λ by

$$\tau_R = [C(\lambda\tau_0)^p\tau_\lambda]^{\gamma/\alpha} \qquad (11)$$

where

$$C = \frac{a}{\gamma} \beta^{p/4} \frac{24\zeta(4)}{\Gamma(5-p)\zeta(4-p)c_2^p} . \qquad (12)$$

In the above equations, ζ is the Riemann zeta function, $c_2 = hc/k$ and the constants α, β and γ are defined by

$$\alpha = \frac{4-p}{2n+p-2}, \quad \beta = \frac{3(n+p/2-1)}{n+1}, \quad \gamma = \frac{2}{1-n}. \qquad (13)$$

The parameter which is still undetermined, namely the constant A in equation (10), is related to the constant a in equation (6) and, following Larson (1969b), this is arbitrarily chosen so that it satisfies the condition that the resulting spectrum must give the correct luminosity L when integrated over all wavelengths. For $n = 3/2$, the values of A determined in this manner for $p = 1$, $3/2$ and 2 are 0.831, 0.645 and 0.354 respectively.

The simple approximation to the temperature distribution as applied in the grey case ($p = 0$, $\tau_R = \tau_\lambda = \tau$), gives

$$T^4 = \left[1 + \frac{3(n-1)}{n+1} \, \tau \right] T_f^4. \tag{14}$$

Substituting equation (8), together with the black body radiation law, into equation (3), we obtain the following expression for L_λ;

$$L_\lambda = C^{-\gamma} \frac{c_1}{\sigma} \frac{T_0 L}{(\lambda T_0)^{5 + \gamma P}} \int_0^\infty \frac{\tau_\lambda^{-\gamma} \, G_n(\tau_\lambda) \, d\tau_\lambda}{\exp \{ c_2 / \lambda T_0 f(\tau_R) \} - 1} \tag{15}$$

where $c_1 = 2\,hc^2$. Since L_λ is dependent upon λ only in combination with T_0 in the factor (λT_0), it is most convenient to calculate $L_\lambda / T_0 L$, as the resulting function is then independent of both L and T_0 and is also normalized to unity when integrated with respect to the independent variable λT_0.

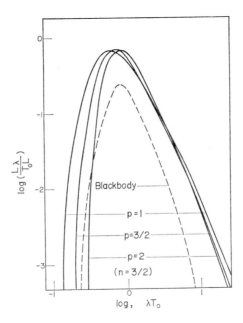

FIG. 66. The emitted spectrum of a protostellar cloud with $n = 3/2$ for $p = 1$, $3/2$ and 2. The black body spectrum is arbitrarily positioned on the diagram for purposes of comparison. (After Larson.)

Spectra which have been calculated numerically from equation (15) for $p = 1$, $3/2$ and 2 are shown in Fig. 66.

The spectra obtained for the grey case are clearly much broader than the black body spectrum for two reasons. First, the shorter wavelength radiation coming from the inner, hotter regions of the cloud is selectively absorbed to a greater extent with increasing p due to the increase in κ_λ towards shorter wavelengths. Second, owing to the reduced opacity and reduced emission at longer wavelengths, there is a decreased amount of long wavelength radiation from the cooler, outer regions. Obviously, the grey case is far from adequate here since these two effects are quite pronounced.

The emitted spectrum of a spherically symmetric protostar may now be calculated during the course of its evolution on the assumption that the opacity of the dust grains follows a

The Nebular Variables

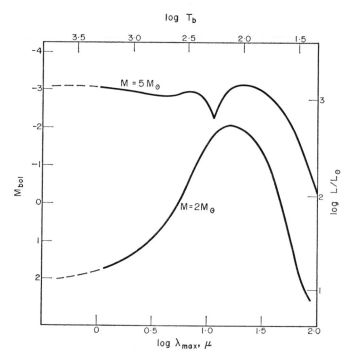

FIG. 67. The evolution is spectral appearance of a protostar for M $= 2\,M_\odot$ and M $= 5\,M_\odot$. (After Larson.)

formula of the type $\kappa_\lambda = \kappa_0\lambda^{-p}$ where suitable values of κ_0 and p may be estimated. The opacity law derived by Larson (1969b) is

$$\kappa_\lambda = 7 \times 10^{-5}\lambda^{-3/2}\ \text{cm}^2\ \text{g}^{-1}. \tag{16}$$

The shape of the emitted spectrum remains constant during the evolution of the protostar and only the parameters L and T_0 vary. T_0 actually determines λ_{max}, the wavelength at which L_λ is a maximum; as T_0 increases, λ_{max} decreases such that the product $\lambda_{max}\,T_0 =$ constant.

During the early stages of their evolution, the protostars for both of the cases we are considering are qualitatively similar, but differ in the later stages since the luminosity in case (b) is now supplied by radiative cooling of the stellar core and no longer by the kinetic energy inflow to the core from the infalling material.

The evolution in spectral appearance for both of these cases is shown in Fig. 67.

The position of the star on the Hertzsprung–Russell diagram at the end of the dynamical collapse phase is given in Fig. 67 by the solid dot. In the case of the 5 $M_\odot$ object, there is no convective "Hayashi" phase as we have seen earlier. When the star first becomes visible, it is found much closer to the main sequence than the lower end of the "Hayashi" track.

The emitted spectrum of a protostar when the infalling cloud is no longer optically thick has also been calculated by Larson (1969b) adopting a stellar core temperature of 5000°K and a dust evaporation temperature of 1000°K. As will be seen from the family of curves given in Fig. 68 which have been calculated for different values of the total optical depth of the surrounding cloud, the emitted spectrum consists of two components.

(a) The visible radiation emitted by the stellar core which is still attenuated to a variable

188

extent by the dust present in the infalling cloud and (b) the thermal infrared emission from the shell of dust around the star.

Two important features must be noted here. First, the gradual emergence of the stellar spectrum is clearly shown in Fig. 68 with a decrease in the total optical depth of the cloud.

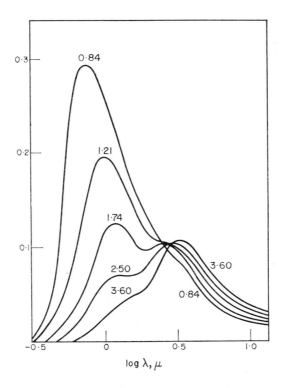

Fig. 68. The emitted spectrum of a protostar when the infalling cloud is no longer optically thick. Core temperature 5000°K and dust evaporation temperature 1000°K. The curves are labelled with the optical depth of the cloud at a wavelength of 1μ. (After Larson.)

Second, the emitted spectrum has been calculated from the equation

$$\frac{L_\lambda}{L} = \exp{(-\tau_\lambda)}\frac{B_\lambda(T_s)}{B(T_s)} + \frac{L_\lambda(\tau_\lambda)}{L} \tag{17}$$

where T_s is the temperature of the black body stellar spectrum and τ_λ is the optical depth at the point where the grains evaporate. The dust emission spectrum $L_\lambda(\tau_\lambda)$ can be obtained from equation (15) with the integral stopped at optical depth τ_λ. It is obvious, therefore, that the resultant spectrum will vary quite appreciably from those illustrated in Fig. 68 if we adopt different values for the black body stellar temperature and also for the dust evaporation temperature.

Here we shall be discussing those objects which have emitted spectra similar to the curves for total optical depths of 2.50 and, particularly, 3.60 where little, if any, of the visible spectrum has emerged. In other words, those in which virtually all of the emitted radiation lies in the infrared.

189

The Nebular Variables

The Becklin–Neugebauer Object

Most of the more comprehensively observed infrared sources of this nature have been found in the Orion Nebula. An infrared point source has recently been discovered by Becklin and Neugebauer (1967) in the Orion complex which, from its general characteristics, appears to be a protostar although the observations may be affected somewhat by interstellar extinction.

The infrared flux for this object shows that it very closely resembles a black body with a temperature of $\sim 700°K$. There is also a close fit with the curve for the emitted spectrum of a protostellar cloud with an opacity given by $\kappa_\lambda \propto \lambda^{-2}$ which is the case for graphite grains. However, the spectrum of this object over the range $2.8–13.5\mu$ obtained by Gillett and Forrest (1973) shows quite strong absorption features at 3.1μ and 10μ.

The feature at 3.1μ has been shown to correspond well with that expected from 10^{-4} g cm^{-2} of H_2O ice grains. The 10μ absorption, on the other hand, is similar in shape to the excess emission observed from several M type stars and requires $\sim 3.5 \times 10^{-4}$ g cm^{-2} of silicate material to reproduce the spectral depth of this feature.

On the assumption that the Becklin–Neugebauer object is a protostar in its early evolutionary stages, it seems that the mass must be appreciably greater than 1 $M_\odot$ and is probably much nearer 5 $M_\odot$. The inferred luminosity of the object is $\sim 10^3 L_\odot$.

The Kleinmann–Low Nebula

A further infrared object which is situated very close to the Becklin–Neugebauer point source has been discovered by Kleinmann and Low (1967). It is evident that this object is much more extended than the former and appears to be a nebula which radiates its peak intensity at longer wavelengths than the Becklin–Neugebauer object.

Gillett and Forrest (1973) have also obtained infrared spectra of this object in the $2.8–13.5\mu$ region. Although the evidence for the presence of the 3.1μ absorption is not as conclusive as in the case of the Becklin–Neugebauer point source, there is a strong absorption at 10μ. Indeed, this is possibly much stronger than that seen against the point source.

Low (1971) has estimated that this infrared nebula has a radius of ~ 0.15 pc, a mass of $\sim 200 M_\odot$ and a total luminosity of $\sim 10^5 L_\odot$. The present picture we have of this object is of a very massive protostar or group of protostars whose energy is provided by a newly formed star (or stars) at the centre of the cloud. Larson (1972) has suggested that the central star might be a main sequence star with a spectral type of about 07 and a mass of $\sim 27 M_\odot$. No H II region is present around this object, an observation which may be explained by the high density of the infalling material (particularly if the star is still accreting matter). As shown by Larson and Starrfield (1971), such an H II region will not form until the star has become more massive and luminous.

Penston *et al.* (1971) have postulated that, since the Becklin–Neugebauer object appears to be peripherally superimposed upon this tenuous, extended nebula (and may be physically associated with it), this point source may be a highly reddened star and responsible for heating the Kleinmann–Low nebula.

Larson (1972) has given reasons why this is not the case. If, as suggested above, the Becklin–Neugebauer object is a main sequence O type star that radiates like a black body in the infrared, then to be the energizing star for the nebula, it would be at least 500 times fainter than it actually is, especially around 5μ. For the infrared brightness of the energizing star to be consistent with the Becklin–Neugebauer object it would have to be of late spectral type, around K0 or later. That this is unlikely is shown by the spectroscopic features found in the spectrum. Penston *et al.*(1971), for example, have shown that molecular absorption features are absent indicating an early spectral class.

We must also examine the age of a main sequence star of this luminosity and the free fall time of the Kleinmann–Low nebula. Here, again, we find an inconsistency. Whereas the former is $\sim 5 \times 10^6$ years, the latter is only $\sim 10^5$ years and there are problems when we attempt to explain how such a nebula could have remained for so long around the star or even how it could form in the neighbourhood of such a luminous star.

There are also grave difficulties if we assume, instead, that the energizing star is a pre-main sequence supergiant. Then its age could not be longer than about 10^3 years and the protostellar cloud from which it formed would have to be extremely dense and have a free fall time not in excess of about the same period.

The available evidence, therefore, points to the Becklin–Neugebauer object being a separate protostar of much smaller mass than the nebula although we cannot rule out the possibility that it is physically connected with it.

Infrared Sources in W3

The galactic radio source W3, lying in the Perseus spiral arm, is closely associated with the nebula IC 1795 and also with two, possibly more, centres of molecular maser emission. It has been known for some years that large quantities of dust and neutral hydrogen are present in this region. According to Reifenstein *et al.* (1970), the kinematic distance to W3 is 3.1 kpc.

Webster and Altenhoff (1970) and Wynn-Williams (1971) have found, from aperture synthesis maps of the region at 2.7 GHz and 5 GHz, that much of the radio emission has its origin in compact condensations which have electron densities of $\sim 10^4$ cm^{-3} and diameters ~ 0.1 pc.

Wynn-Williams *et al.* (1972) have discovered nine infrared objects in W3 by means of high resolution mapping and photometric observations in the spectral range 1.65–20μ. The measurements were made with the Hooker 100-in. and Hale 200-in. reflectors. Since none of these infrared sources are visible optically, it was necessary to establish their positions relative to nearby field stars.

The nine infrared sources discovered by Wynn-Williams *et al.* (1972) may be divided into four groups:

SOURCES ASSOCIATED WITH W3(A)

At 20μ IRS1 (infrared source No. 1) has a shape and size that are virtually identical with those of the radio continuum region W3(A) at 6 cm. The emission, at both wavelengths, seems to be concentrated in an incomplete ring which encloses a central region of lower brightness. Observations at 2.2μ, however, show that in addition there is a very small, unresolved source, IRS2, with a diameter of $\sim 5''$.

Having such a small diameter, it is possible that IRS2 may be the exciting star of W3(A)/IRS1. This belief is further strengthened by the fact that, not only is its infrared flux distribution much bluer than that of the more extended source IRS1, but it also appears to be located very close to the geometric centre of W3(A). The absolute magnitude of IRS2 at 2.2μ is $-5^m.0$ and according to Johnson (1966) its may, therefore, be either an M3 giant

The Nebular Variables

star or an O5 main sequence star. An M3 giant appears unlikely since such a star will produce negligible ionization. An O5 star, however, has sufficient ionizing flux to ionize W3(A) as required on the basis of radio data obtained by Wynn-Williams (1971).

SOURCES ASSOCIATED WITH H II REGIONS

From careful positional measurements, it appears that both IRS3 and IRS4 coincide with the radio sources W3(B) and W3(C) respectively. The former is also clearly extended like W3(B) while the latter has a diameter of less than 5″ (in good agreement with the diameter of W3(C)). There is some evidence that the infrared energy between 3 and 20μ originates from hot dust which is probably mixed with the ionizing gas. Such a situation has been found by Kleinmann (1970) in M17 and by Ney and Allen (1969) in the Orion Nebula.

SOURCES NOT ASSOCIATED WITH H II REGIONS

Three new sources, IRS5, IRS6 and IRS7 were discovered by Wynn-Williams et al. (1972), none of which appear to be associated with any radio or optical feature. Both IRS6 and IRS7 are very faint and have been observed only at 20μ. IRS5, however, is a much more remarkable object since it possesses a very high surface brightness at long wavelengths, is extremely compact, has an infrared spectrum totally unlike that of other sources which have been identified with H II condensation, and is coincident with an H_2O maser source.

As yet, very little is known concerning the nature of this object. It may possibly be a dust-enshrouded evolved supergiant star or a main sequence OB star although it is unlikely that W3, which is a very young HII region, contains an evolved supergiant star.

The alternative possibility is that it is a pre-main sequence protostar. If this is so, then from the models examined by Larson (1969b) its mass must be much greater than $5\,M_\odot$. In the present context, it is perhaps significant that the intrinsic 20μ flux from IRS5 is very like that of the Kleinmann–Low Nebula which Hartmann (1967) has interpreted as a group of protostars.

SOURCES ASSOCIATED WITH W3(OH)

Two small infrared sources, IRS8 and IRS9 are present centered upon W3(OH). The former is the more powerful of the two at 20μ, whereas the latter appears brighter at 2.2μ.

The luminosity and energy distribution of IRS8 are similar to those of IRS3 and IRS4. IRS9 does not appear to be associated with any radio or optical features. A description of this object has been given by Neugebauer et al. (1969).

IRC+10216

The infrared source IRC+10216 in the constellation of Leo is an extended object which is located out of the galactic plane in an unreddened region of the sky. Becklin et al. (1969) have shown that in the region of 5μ it is the brightest source observed outside of the solar system.

At 2.2μ the brightness is variable with an amplitude of up to 2^m and a time scale of the order of 600 days. In this respect, therefore, it shows a resemblance to some of the irregular and long period variables in the optical region. The energy distribution is like that of a black body with a temperature of 650°K. No spectral features were observed by these authors in the wavelength region between 1.5 and 14μ.

Miller (1970) has carried out scanner observations in the near infrared between 0.7 and

1.1μ which show the presence of three absorption features identified as CN bands. Kellermann and Pauliny-Toth (1971) have measured a flux density of 0.31 ± 0.10 f.u. at a wavelength of 3.5μ which is in good agreement with the value to be expected from extrapolation of the infrared measurements. It is generally agreed that this object is a galactic carbon star which is surrounded by an optically thick dust shell, possibly a product of the star itself during its evolution away from the main sequence.

Independent measurements made by Wilson (1971) at 3.5μ are in excellent agreement with those of Kellermann and Pauliny-Toth. The former has also shown that if this radiation comes from an optically thick circumstellar shell, its temperature is $\sim 600^\circ$K and its linear dimension is approximately 100 a.u. The similarity between IRC+10216 and objects such as VY Canis Majoris and NML Cygni is also pointed out. Both of these infrared sources have luminosities of $\sim 10^5 L_\odot$ and it may be significant that VY Canis Majoris is also situated very close to a young stellar cluster on the edge of an HII region. The optical variations of this irregular variable have been described by Florya (1937).

Infrared Objects in Cygnus

A photographic survey for infrared stars in several fields in Cygnus has been carried out by Ackermann (1970) who has reported photoelectric measurements of a sample of extremely red stars in these regions. Some 400 new infrared stars were discovered, all of which are concentrated to the galactic plane. The highly reddened stars are correlated either with dark clouds (Field 95–O) or strong HII regions (Field 78–O). From the observations it appears that some of these infrared sources may represent highly reddened O type stars and possibly the, until now unknown, excitation stars of some HII regions.

Infrared Sources in NGC 2264 and M1-82

A bright infrared source is known in NGC 2264 and several others have been found in M1-82. Allen (1972) has secured direct photographs and low-dispersion slit spectra of these objects in an attempt to determine their exact nature. The former appears to be a heavily reddened star whereas the latter are probably young Be type stars which are obscured by circumstellar envelopes.

Cocoon Stars

The properties of a newly formed massive star have been examined theoretically by Davidson and Harwit (1967) and Davidson (1970). According to this hypothesis, the star is likely to be surrounded by a very dense gas and dust cloud by the time it approaches the main sequence. Radiation pressure, due to the high luminosity of such a star, will force part of the inner material radially outward in the period before the star begins to produce ionizing radiation and this will have a marked effect upon the formation of an HII region in the vicinity of the star.

An exceptionally dense dust front may then precede the eventual ionization front forming a cocoon around the star. In the event that the dust cloud is composed of small graphite grains, extraordinarily large far infrared fluxes are possible. The unusual colours found for some of the stars in the Cygnus association, VI Cygni, by Reddish *et al.* (1967), indicative of the presence of very dense obscuration, may represent observational evidence for the existence of such cocoon stars.

The Nebular Variables

Far Infrared Sources

Broad band measurements have been made by Emerson *et al.* (1973) in the spectral region from 40 to 350μ of a number of infrared sources by means of balloon-carried instruments. During these observations, two new far infrared sources were discovered which are intimately associated with the H II regions RCW 117 ($\alpha = 17^h\ 05^m\ 36^s;\ \delta = -41°\ 32'.4$) and DR 15 ($\alpha = 20^h\ 30^m\ 34^s;\ \delta = +40°\ 04'.4$), both positions given for epoch 1950.0. The radio positions agree very closely with the above infrared positions.

Haro has suggested that RCW 117 may be a planetary nebula and, as such, it appears in the Catalogue of Galactic Planetary Nebulae of Perek and Kahoutek (1967). Rubin and Turner (1971), however, class this object as a compact H II region which agrees with the magnitude of the far infrared flux found by Emerson *et al.* The most recently derived distance of this object is 5 kpc found from studies of the hydrogen 21 cm absorption by Rhadhakrishnan *et al.* (1972). Emerson *et al.* (1973) find that RCW 117 has a luminosity of $16 \times 10^5\ L_\odot$ in the 40 to 350μ region and the total mass of dust present is 0.8 $M_\odot$.

DR 15 possesses an H 109α line emission and radio continuum spectrum which is characteristic of a thermal H II region although this does not appear to be a compact region like RCW 117. It has a lower luminosity $L_{40\text{-}350}$ than the former object of $4 \times 10^5\ L_\odot$ and contains a smaller mass of dust amounting to 0.2 $M_\odot$. The dust temperature in both objects is assumed to be $\sim 80°K$.

In this context it is of interest to note that optical observations of compact H II regions made by Schraml and Mezger (1969) suggest that the ionization of these regions is produced through a dust cloud which surrounds the exciting star or stars.

Earlier observations at balloon altitudes (30 km) had been made by Furniss and his colleagues (1972) using a gallium-doped, germanium bolometer cooled to liquid helium temperatures operating on the UCL balloon-borne telescope of 40 cm aperture. The wavelength limits here were the same as those for the observations made later by Emerson *et al.* (1973).

In this instance, far infrared determinations were made of the nebulae M 42 and NGC 2024 in Orion. The total flux measured for M 42 was 106×10^{-14} W cm^{-1}, in good agreement with the value of $(90 \pm 10) \times 10^{-14}$ W cm^{-1} determined by Low and Aumann (1970).

Harper and Low (1971) have reported the infrared flux from NGC 2024 to be 42×10^{-14} W cm^{-1} in a very wide band extending from 45 to 750μ. As might be anticipated, this is appreciably larger than the value of 25×10^{-14} W cm^{-1} found by Furniss *et al.* in the more restricted range.

Microwave Observations of Infrared Sources

Since the detection of interstellar hydroxyl (OH) molecules by radio astronomers in 1963, a search has been made for the presence of these molecules in various regions including (since 1968) some infrared stars.

The frequencies of these microwave lines are 1612, 1665, 1667 and 1720 MHz. There are three known electronic excited states of the hydroxyl molecule above the ground state, separated from it by $\sim 30,000$, 60,000 and 90,000 cm^{-1}. Since radio emission is often observed from the ground state, this would imply that the hydroxyl molecules are acted upon by far infrared emission.

Certain unique characteristics of the OH emission from infrared stars have been discussed by Wilson and Barrett (1968) and Wilson and Barrett (1970). These are as follows.

(a) The 1912 MHz line is, by far, the strongest in emission and is not polarized.

(b) There is no associated radio continuum found with the OH infrared source.

(c) The profile of the 1612 MHz line shows two distinct velocity peaks which are separated by ~ 40 km/sec. Wilson and his colleagues found that these peaks are only weakly circularly polarized although other observers have reported high polarization for them.

In all cases where OH and H_2O emission have been observed in infrared stars it is highly non-thermal in nature and must therefore be attributed to maser action. The 1612 MHz line observed in NML Cygni is one of the strongest OH emission lines which has been detected at the Earth; the absolute luminosity has been estimated by various workers to be of the order of 4×10^{27} erg sec^{-1}. Some very long baseline studies have been made of this object. These reveal that it contains several small features $\sim 0''.05$ in diameter which suggest a brightness temperature of $\sim 10^{10}$ °K within a small area centered upon this particular source.

It is significant that no OH emission has been found in any of the T Tauri like variables which are strong infrared sources, whereas such emission is present in several of the Mira stars, e.g. R Aquilae, U Orionis, VX Sagittarii and WX Serpentis. It thus appears that a necessary, although not a sufficient, condition for OH emission is the presence of a red giant or supergiant star whose light is variable to some extent.

The first model for this type of object was put forward by Shklovsky (1967) consisting of such a red giant with a stellar wind in which the formation of dust and molecules takes place. This idea was later examined and enlarged upon by Litvak (1969), Litvak et al. (1969), Turner (1969), and Wilson and Barrett (1970).

On this hypothesis, the stellar wind from an expanding, evolved giant or supergiant of late spectral type, having a luminosity of between 10^4 and 10^5 $L_\odot$, takes the form of an extensive, expanding atmosphere in whose outermost regions, a combination of low temperature (~ 600°K) and low density ($\sim 10^{-17}$ g cm^{-3}) allow for the formation of an OH maser. Radiation pressure from the central star forces the stellar material through this atmosphere where turbulence produces discrete regions in which the OH masering originates.

The required rate of mass loss, approximately 10^{-6} $M_\odot$ year^{-1}, is high but probably not excessively so. The OH pumping may be initiated by absorption of a 2.8μ photon which can come either from the star itself or the surrounding dust cloud. Some evidence for this mechanism comes from the observed absence of the 1720 MHz line in these sources. We may postulate that the 2.8μ radiation excites the OH molecules to the first vibrational state, from which they will eventually cascade down and, in conjunction with resonance trapping, will invert the 1612 MHz populations. This will result in the 1720 MHz transition being anti-inverted and consequently it will not appear.

The OH emission which is found associated with H II regions and infrared nebulae has been shown by Raimond and Eliasson (1969) and Neugebauer et al. (1969) to be strong, non-thermal 1665 MHz emission and a different pumping mechanism, possibly collisional, is required to explain this.

Radio Emission from Infrared Objects

A small number of infrared sources have been found to be radio emitters, particularly at frequencies of 1415 and 5000 MHz.

The Nebular Variables

MWC 349

This peculiar, highly reddened Be type star in Cygnus ($\alpha = 20^h\ 30^m\ 56.85^s$; $\delta = +40°29'$ $20''.40$, 1950.0) has been shown by Swings and Struve (1942) to have a strong infrared excess. It appears to be $\sim 10^m$ of obscuration which, according to Ackermann (1970) and Geisel (1970), is almost certainly due to a dense cloud of gas and dust surrounding the star.

The weak radio source close to this star was first reported by Braes *et al.* (1972) with a flux density of 0.060 f.u. at 1415 MHz. Variations in the flux density have been found by Altenhoff and Wendker (1973) who regard this object as a variable radio source, possibly similar to Antares.

On the basis that MWC 349 is associated with the Cygnus OB 2 association, Reddish (1967) has suggested a distance of 2.1 kpc for this star. From observations at 5000 MHz, Baldwin *et al.* (1973) have calculated a diameter of 0.024 pc and an electron density of 3.4×10^4 cm^{-3} for the radio source which corresponds to a mass of ionized gas of 9×10^{-3} $M_\odot$. The central star is apparently of spectral type B0 or perhaps even earlier. From their observations, Braes *et al.* (1972) have concluded that MWC 349 consists of a binary system with a hot and cool component which are immersed in a circumstellar shell of gas and dust. An investigation of this object at 6.63 and 10.52 GHz by Gregory and Seaquist (1973) indicates that the cloud of ionized gas is some 80 times larger than the surrounding dust cloud. It now seems that the system consists of a dense H II region around an early type star of class $\sim$ B0 associated with a more compact dust shell which may surround a cool component. Whether the dust cloud itself is embedded within the H II region is still a matter of conjecture.

RY SCUTI

Radio emission from a source which coincides with the optical position of RY Scuti has been observed by Hughes and Woodsworth (1973) at a frequency of 10.5 GHz with the 46 m telescope of the Algonquin Radio Observatory with a beam width of $2'.8$. A radial velocity of 30 km/sec was obtained by Merrill (1928) from the emission lines in the spectrum of this star and this was later confirmed by Swings and Struve (1940) who also believed that the star represents some stage in the evolution of certain novae, in spite of the fact that Gaposchkin (1937) obtained a light curve of the β Lyrae type with a period of $11^d.1$.

If this is, indeed, the case, then estimates of the total mass of the system indicate this to be $\sim 100\ M_\odot$, making it a binary system with one of the largest combined masses known. On the assumption that the radio emission is thermal in origin, as for most other radio stars, the total mass of the circumstellar shell and its electron density are of the same order as MWC 349.

References

ACKERMANN, G. (1970) *Astron. and Astrophys.* **8**, 315.
ALLEN, D. A. (1972) *Astrophys. Lett.* **12**, 231.
ALTENHOFF, W. J. and WENDKER, H. K. (1973) *Nature* **241**, 37.
BALDWIN, J. E., HARRIS, C. S. and RYLE, M. (1973) *Ibid.* **241**, 38.
BECKLIN, E. E. and NEUGEBAUER, G. (1967) *Astrophys. J.* **147**, 799.
BECKLIN, E. E., FROGEL, J. A., HYLAND, A. R., KRISTIAN, J. and NEUGEBAUER, G. (1969) *Astrophys. J. Lett.* **158**, L137.
BRAES, L. L. E., HABING, H. J. and SCHOEMAKER, A. A. (1972) *IAU Circ.* No. 2450.
COHEN, M. and WOOLF, N. J. (1971) *Astrophys. J.* **169**, 543.

COHEN, M. (1973) *Mon. Not. Roy. astr. Soc.* **161**, 85.
DAVIDSON, K. and HARWIT, M. (1967) *Astrophys. J.* **148**, 443.
DAVIDSON, K. (1970) *Astrophys. and Space Sci.* **6**, 422.
EMERSON, J. P., JENNINGS, R. E. and MOORWOOD, A. F. M. (1973) *Nature Phys. Sci.* **241**, 108.
FELDMAN, P. A., REES, M. J. and WERNER, M. W. (1969) *Nature* **224**, 752.
FLORYA, H. (1937) *Trudy. Sternberg Astr. Inst.* **8**, 2.
FURNISS, I., JENNINGS, R. E. and MOORWOOD, A. F. M. (1972) *Nature Phys. Sci.* **236**, 6.
GAPOSCHKIN, S. (1937) *H.A.* **105**, 509.
GEISEL, S. L. (1970) *Astrophys. J. Lett.* **161**, L105.
GILLETT, F. C., LOW, F. J. and STEIN, W. A. (1968) *Astrophys. J.* **154**, 677.
GILLETT, F. C. and FORREST, N. J. (1973) *Ibid.* **179**, 483.
GREGORY, P. C. and SEAQUIST, E. R. (1973) *Nature Phys. Sci.* **242**, 101.
HARPER, D. A. and LOW, F. J. (1971) *Astrophys. J. Lett.* **165**, L9.
HARTMANN, W. K. (1967) *Ibid.* **149**, L87.
HERBIG, G. H. (1958) *Astrophys. J.* **128**, 259.
HOYLE, F. and WICKRAMASINGHE, N. (1962) *Mon. Not. Roy. astr. Soc.* **124**, 417.
HUGHES, V. A. and WOODSWORTH, A. (1973) *Nature Phys. Sci.* **242**, 116.
JOHNSON, H. L. (1966) *Mon. Not. Roy. astr. Soc.* **145**, 271.
KELLERMANN, K. I., PAULINY-TOTH, I. I. K. (1971) *Astrophys. J.* **166**, 117.
KLEINMANN, D. E. and LOW, F. J. (1967) *Astrophys. J. Lett.* **149**, L1.
KLEINMANN, D. E. (1970) *Bull. Am. Astr. Soc.* **2**, 325.
LARSON, R. B. (1969a) *Mon. Not. Roy. astr. Soc.* **145**, 271.
LARSON, R. B. (1969b) *Ibid.* **145**, 297.
LARSON, R. B. and STARRFIELD, S. (1971) *Astron. and Astrophys.* **13**, 190.
LARSON, R. B. (1972) *Mon. Not. Roy. astr. Soc.* **157**, 121.
LITVAK, M. M. (1969) *Astrophys. J.* **156**, 471.
LITVAK, M. M., ZUCKERMAN, B. and DICKENSON, D. F. (1969) *Astrophys. J. Lett.* **157**, 875.
LOW, F. J. and AUMANN, H. H. (1970) *Ibid.* **162**, L79.
LOW, F. J. (1971) *Dark Nebulae, Globules and Protostars*, p. 115, Univ. of Arizona, Tucson.
MERRILL, P. W. (1928) *Astrophys. J.* **67**, 179.
MILLER, J. S. (1970) *Astrophys. J. Lett.* **161**, L95.
NEUGEBAUER, G. and LEIGHTON, R. B. (1965) *Two Micron Sky Survey: a Preliminary Catalogue*, NASA, SP-3047.
NEUGEBAUER, G., HILGEMAN, T. and BECKLIN, E. E. (1969) *Bull. Am. Astr. Soc.* **1**, 201.
NEY, E. P. and ALLEN, D. A. (1969) *Astrophys. J. Lett.* **155**, L193.
PENSTON, M. V., ALLEN, D. A. and HYLAND, A. R. (1971) *Ibid.* **170**, L33.
PEREK, L. and KAHOUTEK, L. (1967) *Catalogue of Galactic Planetary Nebulae*, Akademia, Prague.
RAIMOND, E and ELIASSON, B. (1969) *Astrophys. J.* **155**, 817.
REDDISH, V. C. (1967) *Mon. Not. Roy. astr. Soc.* **135**, 251.
REDDISH, V. C., LAWRENCE, L. C. and PRATT, N. M. (1967) *Publ. Roy. Obs. Edinburgh*, **5**, 111.
REIFENSTEIN, E. C. III, WILSON, T. L., BURKE, B. F., MEZGER, P. G. and ALTENHOFF, W. J. (1970) *Astron. and Astrophys.* **4**, 357.
RHADHAKRISHNAN, V., GOSS, W. M., MURRAY, J. D. and BROOKS, J. W. (1972) *Astrophys. J.* (Suppl. No. 203) **24**, 49.
RIEKE, G., LEE, T. and COYNE, G. (1972) *Publ. astr. Soc. Pacif.* **84**, 37.
RUBIN, R. H. and TURNER, B. E. (1971) *Astrophys. J.* **165**, 471.
SCHRAML, J. and MEZGER, P. G. (1969) *Ibid.* **156**, 269.
SHKLOVSKY, I. S. (1967) *Astr. Circ. U.S.S.R.* **424**, 1.
SIMON, T., MORRISON, N. D., WOLFF, S. C. and MORRISON, D. (1972) *Astron. and Astrophys.* **20**, 99.
SWINGS, P. and STRUVE, O. (1940) *Astrophys. J.* **91**, 546.
SWINGS, P. and STRUVE, O. (1942) *Ibid.* **95**, 152.
TURNER, B. E. (1969) *Astrophys. J. Lett.* **157**, L103.
WALKER, M. F. (1956) *Astrophys. J. Suppl.* **2**, 365.
WALKER, M. F. (1959) *Astrophys. J.* **130**, 57.
WELIN, G. (1971) *Astron. and Astrophys.* **12**, 312.
WILSON, W. J. and BARRETT, A. H. (1968) *Science* **161**, 778.
WILSON, W. J. and BARRETT, A. H. (1970) *Astrophys. J. Lett.* **160**, 545.
WILSON, W. J. (1971) *Astrophys. J.* **166**, 113.
WOOLF, N. J. and NEY, E. P. (1969) *Astrophys. J. Lett.* **155**, L181.
WYNN-WILLIAMS, C. G. (1971) *Mon. Not. Roy. astr. Soc.* **151**, 397.
WYNN-WILLIAMS, C. G., BECKLIN, E. E. and NEUGEBAUER, G. (1972) *Ibid.* **160**, 1.

Variable Star Index

Variable Star Index

Name Index

Name Index

Name Index

Subject Index

Subject Index

Subject Index

209